Numerical Modeling of Materials under Extreme Conditions

Numerical Modeling of Materials under Extreme Conditions

Editors

Yao Shen
Ning Gao

MDPI • Basel • Beijing • Wuhan • Barcelona • Belgrade • Manchester • Tokyo • Cluj • Tianjin

Editors
Yao Shen
Shanghai Jiao Tong University
China

Ning Gao
Shandong University
China

Editorial Office
MDPI
St. Alban-Anlage 66
4052 Basel, Switzerland

This is a reprint of articles from the Special Issue published online in the open access journal *Metals* (ISSN 2075-4701) (available at: https://www.mdpi.com/journal/metals/special_issues/modeling_extreme_conditions).

For citation purposes, cite each article independently as indicated on the article page online and as indicated below:

LastName, A.A.; LastName, B.B.; LastName, C.C. Article Title. *Journal Name* **Year**, *Volume Number*, Page Range.

ISBN 978-3-0365-7586-5 (Hbk)
ISBN 978-3-0365-7587-2 (PDF)

© 2023 by the authors. Articles in this book are Open Access and distributed under the Creative Commons Attribution (CC BY) license, which allows users to download, copy and build upon published articles, as long as the author and publisher are properly credited, which ensures maximum dissemination and a wider impact of our publications.

The book as a whole is distributed by MDPI under the terms and conditions of the Creative Commons license CC BY-NC-ND.

Contents

About the Editors vii

Yao Shen and Ning Gao
Numerical Modeling of Materials under Extreme Conditions
Reprinted from: *Metals* 2023, 13, 680, doi:10.3390/met13040680 1

Min Sik Lee, Chul Kyu Jin, Junho Suh, Taekyung Lee and Ok Dong Lim
Investigation of Collision Toughness and Energy Distribution for Hot Press Forming Center Pillar Applied with Combination Techniques of Patchwork and Partial Softening Using Side Crash Simulation
Reprinted from: *Metals* 2022, 12, 1941, doi:10.3390/met12111941 5

Wenxue Ma, Yibin Dong, Miaosen Yu, Ziqiang Wang, Yong Liu, Ning Gao, et al.
Evolution of Symmetrical Grain Boundaries under External Strain in Iron Investigated by Molecular Dynamics Method
Reprinted from: *Metals* 2022, 12, 1448, doi:10.3390/met12091448 21

Zigen Xiao, Yun Huang, Zhixiao Liu, Wangyu Hu, Qingtian Wang and Chaowei Hu
The Role of Grain Boundaries in the Corrosion Process of Fe Surface: Insights from ReaxFF Molecular Dynamic Simulations
Reprinted from: *Metals* 2022, 12, 876, doi:10.3390/ met12050876 33

Li Wang, Zhen Wang, Yaping Xia, Yangchun Chen, Zhixiao Liu, Qingqing Wang, et al.
Effects of Point Defects on the Stable Occupation, Diffusion and Nucleation of Xe and Kr in UO_2
Reprinted from: *Metals* 2022, 12, 789, doi:10.3390/met12050789 45

Yaping Xia, Zhen Wang, Li Wang, Yangchun Chen, Zhixiao Liu, Qingqing Wang, et al.
Molecular Dynamics Simulations of Xe Behaviors at the Grain Boundary in UO_2
Reprinted from: *Metals* 2022, 12, 763, doi:10.3390/met12050763 59

Ziqiang Wang, Miaosen Yu, Chen Yang, Xuehao Long, Ning Gao, Zhongwen Yao, et al.
Effect of Radiation Defects on Thermo–Mechanical Properties of UO_2 Investigated by Molecular Dynamics Method
Reprinted from: *Metals* 2022, 12, 761, doi:10.3390/ met12050761 73

Hui Dai, Miaosen Yu, Yibin Dong, Wahyu Setyawan, Ning Gao and Xuelin Wang
Effect of Cr and Al on Elastic Constants of FeCrAl Alloys Investigated by Molecular Dynamics Method
Reprinted from: *Metals* 2022, 12, 558, doi:10.3390/met12040558 85

Linyu Li, Hao Wang, Ke Xu, Bingchen Li, Shuo Jin, Xiao-Chun Li, et al.
Atomic Simulations of the Interaction between a Dislocation Loop and Vacancy-Type Defects in Tungsten
Reprinted from: *Metals* 2022, 12, 368, doi:10.3390/met12030368 97

Zhixiao Liu, Mingyang Ma, Wenfeng Liang and Huiqiu Deng
A Mechanistic Study of Clustering and Diffusion of Molybdenum and Rhenium Atoms in Liquid Sodium
Reprinted from: *Metals* 2021, 11, 1430, doi:10.3390/met11091430 109

Qingqing Zeng, Zhixiao Liu, Wenfeng Liang, Mingyang Ma and Huiqiu Deng
A First-Principles Study on Na and O Adsorption Behaviors on Mo (110) Surface
Reprinted from: *Metals* 2021, 11, 1322, doi:10.3390/met11081322 121

Wenhong Ouyang, Wensheng Lai, Jiahao Li, Jianbo Liu and Baixin Liu
Atomic Simulations of U-Mo under Irradiation: A New Angular Dependent Potential
Reprinted from: *Metals* **2021**, *11*, 1018, doi:10.3390/met11071018 . . . . **133**

Zhengxiong Wang, Jiangyi Luo, Wangwang Kuang, Mingjiang Jin, Guisen Liu, Xuejun Jin and Yao Shen
Strain Rate Effect on the Thermomechanical Behavior of NiTi Shape Memory Alloys: A Literature Review
Reprinted from: *Metals* **2023**, *13*, 58, doi:10.3390/met13010058 . . . . **145**

About the Editors

Yao Shen

Yao Shen, Professor, School of Materials Science and Engineering, Shanghai Jiao Tong University. Interests: micromechanics of materials; plasticity and dislocation theory; mechanical behavior under extreme conditions.

Ning Gao

Ning Gao, Professor, Institute of Frontier and Interdisciplinary Science and Key Laboratory of Particle Physics and Particle Irradiation (MOE), Shandong University. Interests: radiation effects; computer simulations; ion implantation experiments.

Editorial

Numerical Modeling of Materials under Extreme Conditions

Yao Shen [1],* and Ning Gao [2],*

1. School of Materials Science and Engineering, Shanghai Jiao Tong University, Shanghai 200240, China
2. Institute of Frontier and Interdisciplinary Science and Key Laboratory of Particle Physics and Particle Irradiation (MOE), Shandong University, Qingdao 266237, China
* Correspondence: yaoshen@sjtu.edu.cn (Y.S.); ning.gao@sdu.edu.cn (N.G.)

1. Introduction and Scope

Materials used under extreme conditions are important in various industrial and defense fields [1,2]. The performance of these materials critically affects the lifetime of related facilities [1,2]. Thus, it is essential to explore the underlying damage mechanisms of materials under extreme conditions. Considering the difficulty and high cost of the experiments carried out to investigate these mechanisms, numerical modeling of the material response is crucial for study in these fields [3,4]. Until now, although various approaches and models from the atomic scale to the macroscale have been used or developed to simulate the mechanical response and microstructural evolution during the processes [4–8], detailed investigations are still needed to further understand the materials under extreme conditions.

The scope of this Special Issue embraces numerical work on material responses to extreme conditions such as high-speed impact or loading, neutron or ion irradiation, and high-pressure and/or high-temperature environment. Related simulation results based on the first-principle molecular dynamics and finite element methods are reported.

2. Contributions

The Special Issue collects contributions from different research groups by focusing on materials under extreme conditions, at various levels, with different simulation methods. Twelve papers, including eleven research papers and one review paper, have been reviewed and accepted by this Special Issue [9–20].

Ouyang et al. [9] fitted a new angular-dependent potential for a U-Mo system, which well reproduces the macroscopic properties of the system. The threshold displacement energy surface at intermediate and short atomic distances is also more accurately described by this new potential. Furthermore, simulations based on this potential corroborate the negative role of local Mo depletion in the mitigation of irradiation damage and consequent swelling behavior.

Zeng et al. [10] simulated the adsorption and diffusion behaviors of a Na atom on a Mo (110) surface with the presence of Re and O atoms by the first-principles approach. The result shows that the Re alloy atom can strengthen the attractive interactions between Na/O and the Mo substrate, and the existence of a Na or O atom on the Mo surface can slow the Na diffusion by increasing the diffusion barrier. The surface vacancy formation energy results indicate that the dissolution of Mo is a potential corrosion mechanism in the liquid Na environment with O impurities. Furthermore, Liu et al. [11] performed ab initio molecular dynamics simulations to understand the interactions between the Na solvent and Mo or Re solute in the liquid phase. It was found that Mo_2 and Re_2 dimers can be stabilized in liquid Na and a higher temperature leads to a stronger binding force. The Mo species diffuse faster than the Re species and the diffusivity decreases as the cluster size increases.

Li et al. [12] employed molecular static simulations to investigate the interaction between a 1/2 [111] interstitial or vacancy dislocation loop and a vacancy-type defect

including a vacancy, di-vacancy, and vacancy cluster in tungsten (W), through binding energy calculations. Furthermore, the effect of a vacancy cluster on the mobility of the 1/2 [111] interstitial dislocation loop was also explored by the molecular dynamics method. The results indicate that a vacancy cluster can attract the 1/2 [111] interstitial dislocation loop and pin it at low temperatures. At high temperatures, the 1/2 [111] interstitial dislocation loop can move randomly.

Using molecular dynamics methods, Dai et al. [13] calculated the elastic constants C_{11}, C_{12}, C_{44}, bulk modulus, and shear modulus of FeCrAl alloy, one of the candidate materials for accident-tolerant fuel (ATF) cladding in the nuclear power industry. The results show that the concentrations of Al and Cr have different effects on the elastic constants. When the concentration of Al was fixed, a decrease in bulk modulus and shear modulus with increasing Cr content was observed, which is consistent with previous experimental results. The dependence of elastic constants on temperature was also the same as in the experiments. Investigations into the elastic properties of defect-containing alloys have shown that vacancies, void, interstitials, and Cr-rich precipitations have different effects on the elastic properties of FeCrAl alloys.

Wang et al. [14] investigated the influence of radiation defects on the thermo-mechanical properties of UO_2 within 600–1500 K through the molecular dynamics method. The results indicate that these point defects reduce the thermal expansion coefficient (α) at all studied temperatures. The elastic modulus at finite temperatures decreases linearly with an increase in the concentration of Frenkel defects and antisites. All these results indicate that Frenkel pairs and antisite defects could degrade the performance of UO_2.

Xia et al. [15] simulated the behavior of xenon (Xe) bubbles in uranium dioxide (UO_2) grain boundaries by using the molecular dynamics method. The results indicate that the formation energy of Xe clusters at the $\Sigma 5$ grain boundaries (GBs) is much lower than in the bulk. The diffusion activation energy of a single interstitial Xe atom at the GBs was approximately 1 eV, lower than that in the bulk. The bubble pressure dropped with increasing temperature at low Xe concentrations, whereas the volume increased. Xe atoms were more regular in the bulk, whereas multiple Xe atoms formed a planar structure at the GBs.

Wang et al. [16] studied the effects of point defects on the behaviors of Xe/Kr clusters in UO_2 by using molecular dynamics. The results show that Xe and Kr clusters occupy vacancies as nucleation points by squeezing U atoms out of the lattice, and the existence of vacancies increases the stability of the clusters. Higher temperature and higher concentrations of interstitial Xe/Kr atoms or vacancies in the system all facilitate the formation of the clusters. The activation energy of interstitial Xe/Kr atoms and clusters in UO_2 is ~2 eV, indicating that the diffusion of the interstitial atoms is very difficult.

Xiao et al. [17] simulated the influence of grain boundaries on Fe-H_2O interfacial corrosion through the molecular dynamics method with a new Fe-H_2O reaction force field potential. The results indicate that the corrosion rate at the polycrystalline grain boundary is significantly higher than that of twin crystals and single crystals. By the analysis of stress, it can be found that the stress at the polycrystalline grain boundary and the $\Sigma 5$ twin grain boundary decreases sharply during the corrosion process. The severe stress release at the grain boundary could promote the dissolution of Fe atoms. The formation of vacancies on the Fe matrix surface will accelerate the diffusion of oxygen atoms, resulting in the occurrence of intergranular corrosion.

Ma et al. [18] investigated the evolution of atomic structures and related changes in the energy state, atomic displacement, and free volume of several symmetrical grain boundaries (GB) under the effects of external strain in body-centered cubic (bcc) iron by the molecular dynamics (MD) method. The results indicate that under external strain, two mechanisms are responsible for the failure of these GBs, including slip system activation, dislocation nucleation, and dislocation network formation. This is induced directly by either the external strain field or by phase transformation from the initial bcc to fcc structure under the effects of external strain.

Lee et al. [19] performed a side crash simulation and investigated the effect of hot press forming (HPF) a center pillar with a combination of patchwork (PW) and partial softening (PS) techniques on collision toughness and energy distribution flow. The roles of the PW and PS techniques were verified during the side crashes. PW improves the strain energy and intrusion displacement by 10% and 7.5%, respectively, and PS improves the plastic deformation energy and intrusion displacement by 10%. When PW and PS were applied to the HPF center pillar simultaneously, a synergistic effect was achieved.

Wang et al. [20] reviewed the experiments and models for the effect of strain rate on NiTi shape memory alloys (SMAs). Experimental observations on the rate-dependent properties, such as stress responses, temperature evolutions, and phase nucleation and propagation, under uniaxial loads, are classified and summarized based on the order of the strain rate magnitudes, with the influences of the microstructure on the strain-rate responses briefly discussed. This review of modeling for the rate-dependent behaviors of NiTi SMAs focuses on how the physical origins are reflected or realized in the constitutive relationship.

Acknowledgments: As Guest Editors, we would like to especially thank all the contributing authors, reviewers, and editors.

Conflicts of Interest: The authors declare no conflict of interest.

References

1. Zinkle, S.J.; Was, G.S. Materials challenges in nuclear energy. *Acta Mater.* **2013**, *61*, 735. [CrossRef]
2. Rapp, B. Materials for extreme environments. *Mater. Today* **2006**, *9*, 6. [CrossRef]
3. Zinkle, S.J. Fusion materials science: Overview of challenges and recent progress. *Phys. Plasmas* **2005**, *12*, 058101. [CrossRef]
4. Nordlund, K. Historical review of computer simulation of radiation effects in materials. *J. Nucl. Mater.* **2019**, *520*, 273. [CrossRef]
5. Gao, N.; Yao, Z.W.; Lu, G.H.; Deng, H.Q.; Gao, F. Mechanisms for <100> interstitial dislocation loops to diffuse in BCC iron. *Nat. Commun.* **2021**, *12*, 225. [PubMed]
6. Yang, L.; Wirth, B.D. An improved xenon equation of state for nanobubbles in UO_2. *J. Nucl. Mater.* **2022**, *572*, 154089. [CrossRef]
7. Germann, T.C.; Holian, B.L.; Lomdahl, P.S.; Ravelo, R. Orientation dependence in molecular dynamics simulations of shocked single crystals. *Phys. Rev. Lett.* **2000**, *84*, 5351. [CrossRef] [PubMed]
8. Yin, Y.; Bertin, N.; Wang, Y.; Bao, Z.; Cai, W. Topological origin of strain induced damage of multi-network elastomers by bond breaking. *Extrem. Mech. Lett.* **2020**, *40*, 100883. [CrossRef]
9. Ouyang, W.; Lai, W.; Li, J.; Liu, J.; Liu, B. Atomic Simulations of U-Mo under Irradiation: A New Angular Dependent Potential. *Metals* **2021**, *11*, 1018. [CrossRef]
10. Zeng, Q.; Liu, Z.; Liang, W.; Ma, M.; Deng, H. A First-Principles Study on Na and O Adsorption Behaviors on Mo (110) Surface. *Metals* **2021**, *11*, 1322. [CrossRef]
11. Liu, Z.; Ma, M.; Liang, W.; Deng, H. A Mechanistic Study of Clustering and Diffusion of Molybdenum and Rhenium Atoms in Liquid Sodium. *Metals* **2021**, *11*, 1430. [CrossRef]
12. Li, L.; Wang, H.; Xu, K.; Li, B.; Jin, S.; Li, X.-C.; Shu, X.; Liang, L.; Lu, G.-H. Atomic Simulations of the Interaction between a Dislocation Loop and Vacancy-Type Defects in Tungsten. *Metals* **2022**, *12*, 368. [CrossRef]
13. Dai, H.; Yu, M.; Dong, Y.; Setyawan, W.; Gao, N.; Wang, X. Effect of Cr and Al on Elastic Constants of FeCrAl Alloys Investigated by Molecular Dynamics Method. *Metals* **2022**, *12*, 558. [CrossRef]
14. Wang, Z.; Yu, M.; Yang, C.; Long, X.; Gao, N.; Yao, Z.; Dong, L.; Wang, X. Effect of Radiation Defects on Thermo–Mechanical Properties of UO_2 Investigated by Molecular Dynamics Method. *Metals* **2022**, *12*, 761. [CrossRef]
15. Xia, Y.; Wang, Z.; Wang, L.; Chen, Y.; Liu, Z.; Wang, Q.; Wu, L.; Deng, H. Molecular Dynamics Simulations of Xe Behaviors at the Grain Boundary in UO_2. *Metals* **2022**, *12*, 763. [CrossRef]
16. Wang, L.; Wang, Z.; Xia, Y.; Chen, Y.; Liu, Z.; Wang, Q.; Wu, L.; Hu, W.; Deng, H. Effects of Point Defects on the Stable Occupation, Diffusion and Nucleation of Xe and Kr in UO_2. *Metals* **2022**, *12*, 789. [CrossRef]
17. Xiao, Z.; Huang, Y.; Liu, Z.; Hu, W.; Wang, Q.; Hu, C. The Role of Grain Boundaries in the Corrosion Process of Fe Surface: Insights from ReaxFF Molecular Dynamic Simulations. *Metals* **2022**, *12*, 876. [CrossRef]
18. Ma, W.; Dong, Y.; Yu, M.; Wang, Z.; Liu, Y.; Gao, N.; Dong, L.; Wang, X. Evolution of Symmetrical Grain Boundaries under External Strain in Iron Investigated by Molecular Dynamics Method. *Metals* **2022**, *12*, 1448. [CrossRef]

19. Lee, M.S.; Jin, C.K.; Suh, J.; Lee, T.; Lim, O.D. Investigation of Collision Toughness and Energy Distribution for Hot Press Forming Center Pillar Applied with Combination Techniques of Patchwork and Partial Softening Using Side Crash Simulation. *Metals* **2022**, *12*, 1941. [CrossRef]
20. Wang, Z.; Luo, J.; Kuang, W.; Jin, M.; Liu, G.; Jin, X.; Shen, Y. Strain Rate Effect on the Thermomechanical Behavior of NiTi Shape Memory Alloys: A Literature Review. *Metals* **2023**, *13*, 58. [CrossRef]

Disclaimer/Publisher's Note: The statements, opinions and data contained in all publications are solely those of the individual author(s) and contributor(s) and not of MDPI and/or the editor(s). MDPI and/or the editor(s) disclaim responsibility for any injury to people or property resulting from any ideas, methods, instructions or products referred to in the content.

Article

Investigation of Collision Toughness and Energy Distribution for Hot Press Forming Center Pillar Applied with Combination Techniques of Patchwork and Partial Softening Using Side Crash Simulation

Min Sik Lee [1], Chul Kyu Jin [2,*], Junho Suh [1], Taekyung Lee [1] and Ok Dong Lim [3]

[1] School of Mechanical Engineering, Pusan National University, 2 Busandaehak-ro 63beon-gil, Geumjeong-gu, Busan 46241, Korea
[2] School of Mechanical Engineering, Kyungnam University, 7 Kyungnamdaehak-ro, Masanhappo-gu, Changwon-si 51767, Korea
[3] Autogen Co., Ltd., 249, Sihwa Venture-ro, Siheung-si 15118, Korea
* Correspondence: cool3243@kyungnam.ac.kr; Tel.: +82-55-249-2346

Abstract: Various techniques can be applied to center pillars to enhance collision characteristics during side crashes. For instance, patchwork (PW) can be welded to the center pillar to increase its stiffness, and partial softening (PS) can be applied to provide ductility. Side crash tests are conducted by the Insurance Institute for Highway Safety (IIHS) to evaluate collision resistance. However, it is difficult to evaluate collision toughness and energy distribution flow for each automobile component. In this study, a side crash simulation was performed with IIHS instruction. We investigated the effect of hot press forming (HPF) a center pillar with a combination of PW and PS techniques on collision toughness and energy distribution flow. As a result, the role of PW and PS techniques were verified during side crashes. PW improved the strain energy and intrusion displacement by 10% and 7.5%, respectively, and PS improved the plastic deformation energy and intrusion displacement by 10%. When PW and PS were applied to the HPF center pillar simultaneously, a synergistic effect was achieved.

Keywords: center pillar; side crash simulation; patchwork; partial softening; energy distribution

1. Introduction

In light of global environmental concerns, the automobile industry has become increasingly interested in fuel efficiency and weight reduction. With increasing demand for technological development to achieve high performance and high fuel efficiency, research on the application of advanced high-strength steel (AHSS) for automobile components has progressed [1–4].

The HPF process was introduced to enhance the stiffness and strength of automobile components. In the HPF process, a blank is heated to a high temperature to create an austenitic phase and then cooled rapidly in forming dies to form a martensite phase [5,6].

However, the use of hot stamped parts is restricted in automobile components and requiring collision absorption, owing to their low elongation. For example, in the case of a side crash, the B-pillar, also known as the center pillar, is among the most important automobile components with respect to passenger safety. During side crash impact, the center pillar must be ductile to absorb the collision energy and stiff to improve intrusion resistance. Much research has been conducted on center pillars made with alternative materials [7–9] or post tempering processes to obtain tailored properties [10,11].

Owing to productivity demands and manufacturing costs, the PW and PS techniques are well-known and widely used in the automobile industry to improve the collision characteristics of center pillars.

To improve intrusion resistance and increase the stiffness of automobile components, a PW is attached to the center pillar to achieve reinforced structural stiffness. The formability of laser-welded PW blanks was investigated by Mori et al. and Shi et al. [12,13]. The spot weld conditions to attach PW to HPF parts was optimized by Ahmad et al. [14]. Chengxi et al. studied methods to improve the mechanical properties of B-pillars with PW and predict the temperature distribution using FE simulation [15].

PS can be applied to improve energy absorption and increase the ductility of automobile components. Gao et al. investigated the characteristics of temperature-dependent IHTC of 22MnB5 during the spray-quenching process [16]. Ota et al. [17] evaluated the damage value and tailored temperature using a forming limit diagram. Bok et al. performed a simulation to predict the microstructure and mechanical properties of a B-pillar in comparison with experimental results [18]. Kim et al. manufactured a B-pillar using the partial strengthening method with minimal shape change [19]. However, the above studies investigated only the manufacturing process and were limited in that they did not evaluate the impact toughness and energy distribution flow, which affect performance when applied to an actual vehicle. A few studies have been conducted involving impact tests to evaluate the effects of PW and PS on impact toughness and energy distribution. Recently, M. S. Lee et al. manufactured an HPF center pillar with PW and PS by controlling the cooling rate and conducted experimental and numerical drop weight tests [20]. Owing to the limitations of the experimental equipment, the results were limited in terms of evaluating the energy absorption characteristics of PW and PS, and test collisions could not be conducted at high speed or using real vehicles owing to high costs.

In this study, given the importance of the evaluation of safety in high-speed collisions, a vehicle and an impactor were modeled according to the IIHS guidelines to evaluate the collision absorption capacity at high speed based on a previously verified low-speed drop weight test, as well as energy distribution in a high-speed collision. To improve stiffness and ductility, PW and PS techniques were applied to the HPF center pillar, and experiments were performed for comparison with the simulations to verify the results.

2. Experiment and Simulation

2.1. Materials

Figure 1a shows that the function of the center pillar during a side crash. Generally, the HPF center pillar has a low elongation of less than 5%, so collision toughness is not satisfied. In addition, the stiffness of the center pillar in the top region is not sufficient against intrusion resistance. The PS technique provides ductility to the HPF center pillar and, PW strengthens the stiffness of the HPF center pillar to improve collision characteristics, such as energy absorption and intrusion resistance.

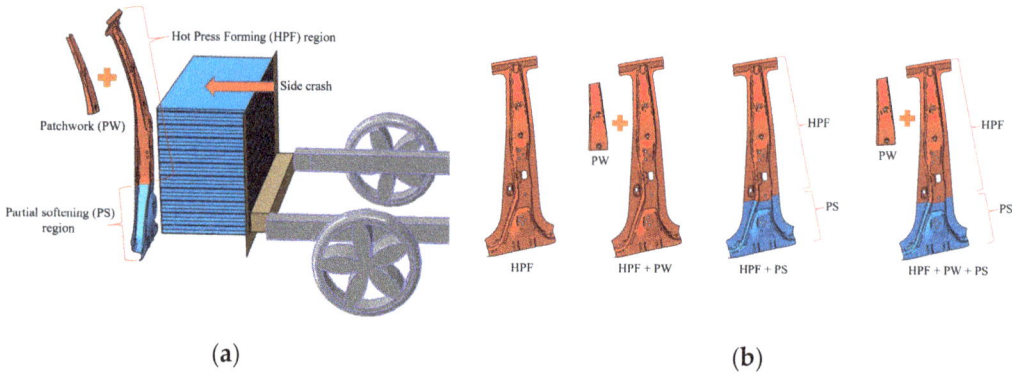

Figure 1. (a) Function of the center pillar during a side crash. (b) Four kinds center pillars [20].

In this study, a 22MnB5 boron steel sheet was used for the HPF center pillar and PW. The PS region was controlled by a cartridge heater during manufacturing of the HPF center pillar [20]. Table 1 shows the material properties of 22MnB5 (HPF) and 22MnB5 (PS). The thicknesses of the HPF center pillar and PW were both 1.2 mm. To evaluate the effect of PW and PS on energy distribution and intrusion resistance, a side crash simulation was performed with four kinds of the center pillar, as shown in Figure 1b.

Table 1. Material properties of 22MnB5 (HPF) and 22MnB5 (PS) [20].

Material	Tensile Strength (MPa)	Elongation (%)	Elastic Modulus (GPa)	Vickers Hardness (HV)
22MnB5 (HPF)	1500	5	210	440
22MnB5 (PS)	715	11~13	210	225

2.2. Drop Weight Test and Simulation

2.2.1. Geometry Modeling, Mesh, and Weld Constraints for FE Simulation

An FE simulation for the drop weight test was performed using ABAQUS/explicit software. Figure 2 shows the geometry modeling and welding constraints between the HPF center pillar and PW. The fine mesh was defined in the impact region, and the rough mesh was defined in other regions to reduce computation time. The element types of the center pillar and weld nodes were S4 and CNN3D2, respectively. The PW was welded to the HPF center pillar with 1200 × 440 × 51.5 mm³ dimension. As shown in Figure 2b, the master weld node was defined on the element of center pillar, and the slave weld node was defined on the element of the PW. The master and slave weld nodes were bonded, and the attachment method used for the welding connector was point to point. The damage criterion and damage evolution of weld constraints were not applied because a fracture did not occur at the weld point. The detailed weld conditions for the FE simulation are listed in Table 2.

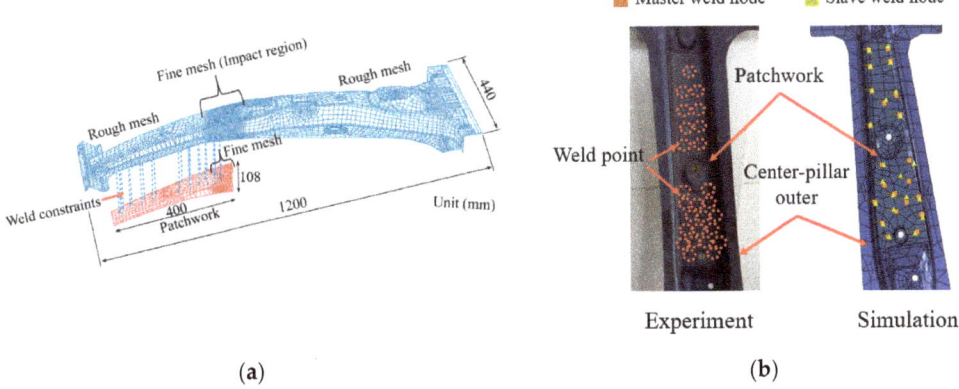

Figure 2. Center pillar and PW: (**a**) geometry and mesh; (**b**) weld constraints.

Table 2. Spot weld conditions for FE simulation.

Attachment Method	Additional Mass (kg)	Spot Radius (mm)	Degrees of Freedom
Point-to-point	0	3	0

2.2.2. Boundary Conditions for Drop Weight Test

Figure 3 shows the simulation and experimental apparatus for the drop weight test. The center pillar was fixed by clamp, as shown in Figure 3a. The drop height from the center pillar was 610 mm, and the load cell was attached to an impactor with a weight of 160 kg.

A drop weight simulation was also performed for comparison with the experimental data. The impactor velocity was 3.450 m/s before collision. The 6 degrees of freedom of the center pillar were fixed in the zig region by a clamp. The friction coefficient between the center pillar and impactor was 0.1. The collision time was 0.18 s. The detailed boundary conditions for the drop weight test are listed in Table 3.

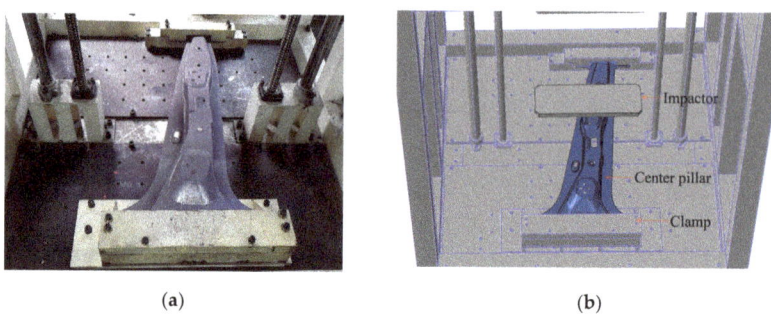

Figure 3. Boundary conditions for the drop weight test: (a) experiment; (b) simulation.

Table 3. Boundary conditions of the simulation for the drop weight test.

Impactor Mass (kg)	Collision Time (s)	Impactor Velocity (m/s)	Drop Height (mm)	Initial Potential Energy (J)	Friction Coefficient between Front Bumper Beam and Impactor
160	0.18	3.45	610	956	0.1

2.3. Side Crash Simulation

2.3.1. Geometry and Properties of a Moving Deformable Barrier in a Sedan

A moving deformable barrier (MDB) was modeled for side crash simulation according to IIHS guidelines. Figure 4 shows the geometric modeling of an MDB with dimensions of $450 \times 1000 \times 650$ mm^3. The MDB consists of two parts: a main honeycomb block and an aluminum sheet used as the cover layer. The two parts were adhesively bonded together. The main honeycomb block was hexagonal in shape with a cell size of 22 mm made Al5052 foil. The foil thickness was 0.05 mm, with a crush strength of 31 MPa [21]. The weight of the MDB was 1360 kg $\pm$ 5 kg with a deformable element, and the center of mass was 1000 mm from the front of the deformable element, as shown in Figure 4. The MDB roll (I_x), pitch (I_y), and yaw (I_z) moments of inertia were 542 kg·m^2, 2471 kg·m^2, and 2757 kg·m^2, respectively.

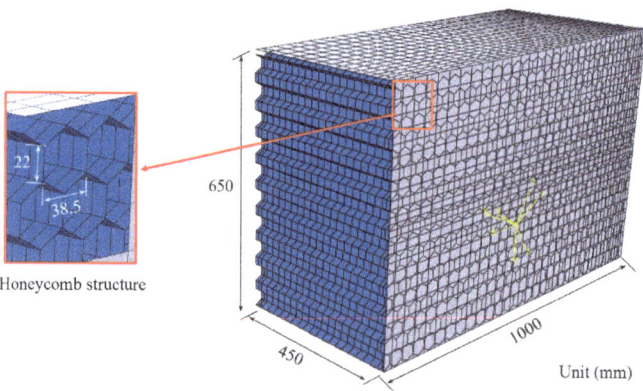

Figure 4. Geometric modeling of an MDB for side crash simulation.

Figure 5 shows the geometric modeling and material section of the automobile and pillars. The automobile geometry for used for simulation had the same dimensions as an automobile: 4580 × 1820 × 1250 mm³. As an actual automobile, the shell model was used, and the element type was set to S4 to reduce the hourglass effect and enable detailed measurement of absorbed energy. Owing to the complexity of the shape of the automobile, with curves and had many holes, the automobile part was meshed using a bottom-up mesh technique. The collision region was finely meshed, whereas regions away from the collision area were roughly meshed. The thicknesses of the center pillar, outer and inner reinforcement, and doors and other frames were 1.2 mm, 1.2 mm, and 2.0 mm, respectively. Multiple materials were used to design the sedan. The center pillar, PW, and frame and other parts were made with HPF with PS, HPF, and DP590 and CR420 (mild steel), respectively.

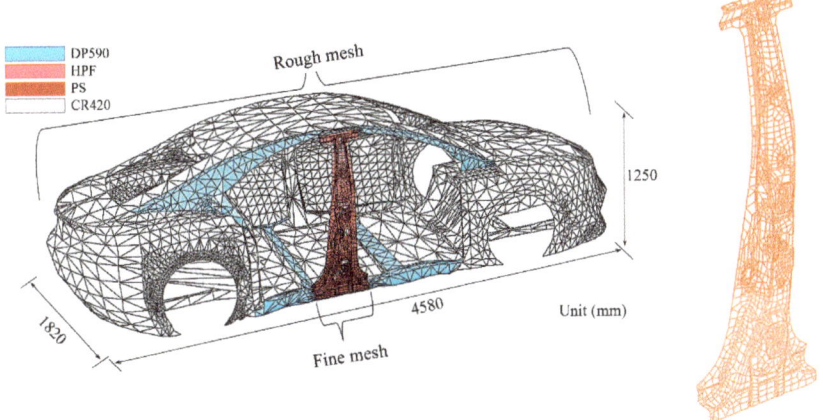

Figure 5. Geometry and material sections of the sedan for side crash test.

2.3.2. Boundary Conditions for Side Crash Test

Figure 6 shows a schematic diagram of the side crash test. A side crash test involves a stationary vehicle, evaluating the impact characteristics when an impactor equipped with an MDB collides with the side region of a stopped vehicle according to IIHS regulation. An impactor with an MDB has a collision velocity of 50 km/h (13.8 m/s) and hits the stopped vehicle at a 90 degree angle, as shown in Figure 6. The mass of the sedan and the impactor with an MDB were 1420 and 1360 kg, respectively. The 6 degrees of freedom of the stopped vehicle were fixed on the ground. The friction coefficient between the ground and the impactor's wheel was 0.1. The collision time was 0.1 s. Because the opposite side of the collision zone would not be deformed, a half body of a stopped vehicle (sedan) was used as a rigid body in order to reduce the analysis time. The detailed side crash conditions are listed in Table 4. Mass scaling was not used, and the stable time increment size was 1.67821×10^{-8} s.

Table 4. Collision conditions for side crash simulation.

Sedan Mass (kg)	MDB Mass (kg)	Collision Time (s)	Impactor Velocity (m/s)	Initial Kinetic Energy (kJ)	Friction Coefficient between Ground and Vehicle
1420	1500	0.1	13.8	142	0.1

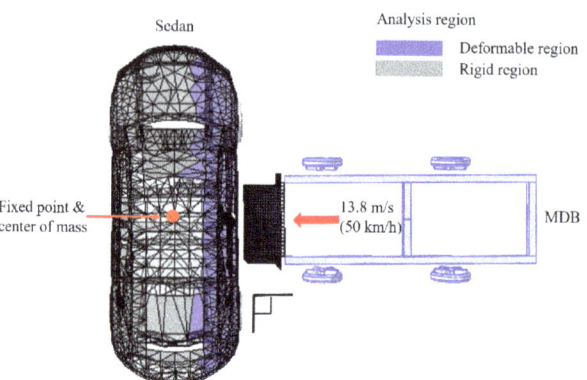

Figure 6. Boundary conditions for side crash simulation.

2.3.3. Energy Absorption Method

In the side crash simulation, the various energies, such as elastic, plastic, kinetic, friction, and viscoelastic energy, were transformed by deformation during the collision. The energy absorbed by deformation was experimentally calculated using load-displacement data according to Equation (1) [22].

$$E_{ab} = \int_0^s F(x)dx \quad (1)$$

where E_{ab}, s, and F are absorbed energy, crash displacement, and impulsive force, respectively.

The specific energy absorption (SEA) was obtained by Equation (2):

$$\text{SEA} = \frac{E_a}{M} \quad (2)$$

where M and E_a are total mass and absorbed energy, respectively.

When the SEA value is high, the energy absorption capability is high. However, this is approximate (rather than exact) estimation method and cannot determine elastic and plastic deformation energy.

In this study, the integrated stress–strain curve data per unit element volume were used to calculate the accurate absorbed energy, such as elastic and plastic dissipated energies, and for analysis of the distribution of energy flow in the time domain.

Figure 7a shows the meshed shell with the S4 element type used to accurately calculate the energy variable data and avoid the hourglass effect. As shown in Figure 7a, the shell element has 4 points, which are inside the mid-surface, in contrast with a solid element type. Because the mid-surface was used for analysis, the change in shell thickness was not visually expressed, and the strain in the thickness direction of the shell element was calculated as Equation (3):

$$\varepsilon_{33} = \frac{\nu}{1-\nu}(\varepsilon_{11} + \varepsilon_{22}) \quad (3)$$

Treating these as logarithmic strains,

$$\ln\frac{t}{t_0} = -\frac{\nu}{1-\nu}\ln\left(\frac{l_0}{l}\right) = -\frac{\nu}{1-\nu}\ln\left(\frac{A}{A^0}\right) \quad (4)$$

where l, t, ν, and A are the element length, thickness, Poisson's ratio, and area of the shell's reference surface, respectively.

The change in shell thickness is expressed by Equation (5).

$$\frac{t}{t_0} = \left(\frac{A}{A_0}\right)^{\left(-\frac{v}{1-v}\right)} \tag{5}$$

According to the above equations, the element volume of each mesh can be measured by Equation (6) after deformation.

$$V_{element} = At_0 \left(\frac{A}{A_0}\right)^{\left(-\frac{v}{1-v}\right)} \tag{6}$$

The stress–strain curve describing the calculation of absorbed energy per element volume during deformation is shown in Figure 7b.

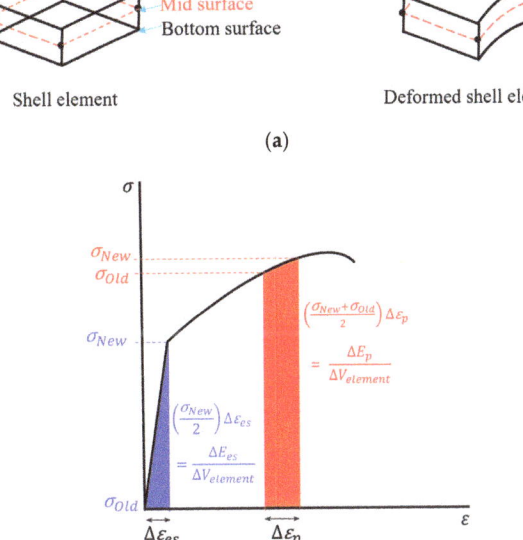

Figure 7. Shell element and energy absorption of the stress–strain curve per element volume: (**a**) element deformation; (**b**) elastic strain energy and plastic deformation energy.

As shown in Figure 7b, the specific elastic and plastic deformation energies are measured by Equations (7) and (8), respectively:

$$\Delta E_{es} = \frac{1}{2}\sigma_{New}\Delta\varepsilon_{es}\Delta V \tag{7}$$

$$\Delta E_p = \frac{1}{2}(\sigma_{Old} + \sigma_{New})\Delta\varepsilon_p \Delta V \tag{8}$$

where σ_{New}, σ_{Old}, σ_u, and ΔV are the new stress, previous stress, user-defined equation stress, and specific element volume, respectively.

The internal energy can be expressed by integrating Equation (9):

$$E_I = \int_0^t \left(\int_V \sigma^u : \dot{\varepsilon}\, dV\right) dt \tag{9}$$

Then, $\dot{\varepsilon} = \dot{\varepsilon}^{es} + \dot{\varepsilon}^p + \dot{\varepsilon}^c$ and E_I can be separated as in Equation (10):

$$E_I = \int_0^t \left(\int_V \sigma^u : \dot{\varepsilon} dV \right) dt = \int_0^t \left(\int_V \sigma^u : \dot{\varepsilon}^{es} dV \right) dt + \int_0^t \left(\int_V \sigma^u : \dot{\varepsilon}^p dV \right) dt + \int_0^t \left(\int_V \sigma^u : \dot{\varepsilon}^c dV \right) d = E_{es} + E_p + E_c \quad (10)$$

where $\dot{\varepsilon}^{es}$, $\dot{\varepsilon}^p$, and $\dot{\varepsilon}^c$ are the elastic strain rate, plastic strain rate, and creep strain rate, respectively; and E_{es}, E_p, and E_c are the elastic energy, plastic energy, and creep strain energy, respectively.

The elastic strain energy (E_{es}) results from linear deformation, whereas energy is dissipated by plasticity (E_p) when permanent deformation of the meshed element begins. The elastic and plastic strain regions are linear and non-linear, respectively. The aforementioned energies are defined by Equations (11) and (12), respectively:

$$E_{es} = \int_0^t \left(\int_V \sigma^u : \dot{\varepsilon}^{es} dV \right) dt = \sum_t \sum_{i=1}^n \frac{1}{2} \sigma_i \varepsilon_i \Delta V_i \quad (11)$$

$$E_p = \int_0^t \left(\int_V \sigma^u : \dot{\varepsilon}^p dV \right) dt \quad (12)$$

In this study, the element set of PW, PS, and MDB was selected, as shown in Figure 8. The E_{es} and E_p of the four kinds of center pillars were compared during a side crash because the elastic strain and plastic deformation energies account for most of the internal energy, which is known as the absorbed energy during a collision. E_{es} and E_p of PS and PW were calculated using the above equation during the deformation of the element according to the four types of center pillars, as shown in Figure 1b.

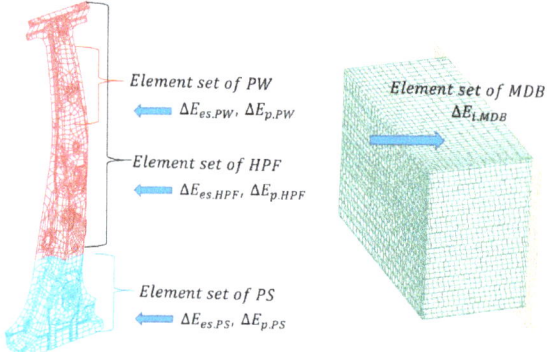

Figure 8. Energy distribution of the center pillar and MDB.

2.3.4. Damage Initiation Criteria and Damage Evolution

Local necking in the shell element could not be realized because the sheet metal in the simulation was very thin. To predict the onset of necking instability, the forming limit diagram (FLD) curve was used in this study. Simulations were performed by applying the previously obtained FLD for 22MnB5 (HPF and PS) to Abaqus/explicit [23,24].

3. Results and Discussion

3.1. Comparison of Impulse between Experiment and Simulation

Figure 9 compares the experimental and simulation results after the collision test for the HPF center pillar with PW and PS. Figure 9a,b show the top and side views of the center pillar, respectively, between the experiment and the simulation. The figure shows that the deformation behaviors of the experiment and the simulation were similar [20].

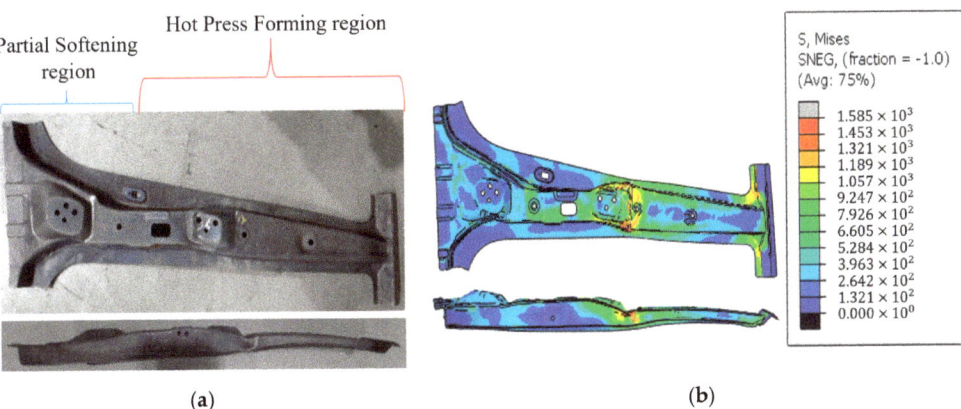

Figure 9. Comparison of the HPF center pillar with PW and PS after the drop weight test: (a) top and side view after the experiment; (b) top and side view after the simulation [20].

Figure 10 shows the comparison result of impactor load between experiment and simulation according to time. The maximum impact load of the experiment and simulation were 9.5 kN and 12.7 kN, respectively. The value of impulse and collision time between the experiment and simulation were 799 N·s, 0.165 s and 787 N·s, 0.117 s, respectively. Based on above results, the simulation and experiment were validated, and these material sections of mechanical properties were used for side crash simulation.

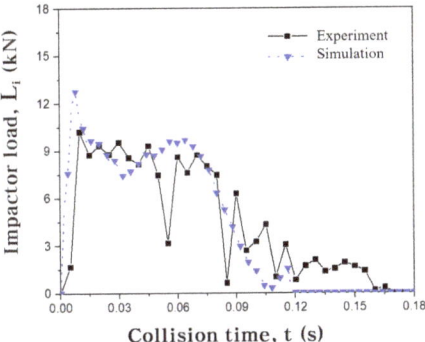

Figure 10. Comparison of the impactor load of the HPF center pillar with PW between the experiment and the simulation [20].

3.2. Side Crash Simulation

3.2.1. Energy Distribution during Side Crash Simulation

The first law of thermodynamics is the law of conservation of energy. In an isolated system, the total energy remains constant, even when the energy changes from one form to another. The energy can be converted into various types in an automobile side impact, such as kinetic energy, potential energy, internal energy, or friction dissipation energy. In general, most kinetic energy in an automobile crash causes a considerable change in internal energy as a result of elastic and plastic deformation. Total energy can be measured as the sum of several energies according to Equation (13):

$$E_{total} = E_i + E_k + E_v + E_f + E_e \qquad (13)$$

where E_i, E_k, E_v, E_f, and E_e are internal energy, kinetic energy, viscous dissipation energy, friction dissipation energy, and external work, respectively.

Figure 11 shows the distribution of the total energy during the side crash. The total energy is 142 kJ. First, the velocity of MDB decreased gradually as a result of friction between the ground and MDB before the collision. The E_k decreased sharply during the collision because the velocity of the MDB decreased, whereas the E_i was increased by plastic and elastic deformation, and E_v increased to 5.8 kJ as a result of material damping. During the collision, other energies increased, but total energies were maintained. For example, when the collision time was 0.0385 s, the $E_{i\cdot MDB}$, $E_{i\cdot Sedan}$, E_k, E_v, and E_f were 45.1 kJ, 45.7 kJ, 45.4 kJ, 2.9 kJ, and 5.9 kJ, respectively, as shown in Figure 11a,b. The value of E_e was 0 kJ because the external work had not yet occurred.

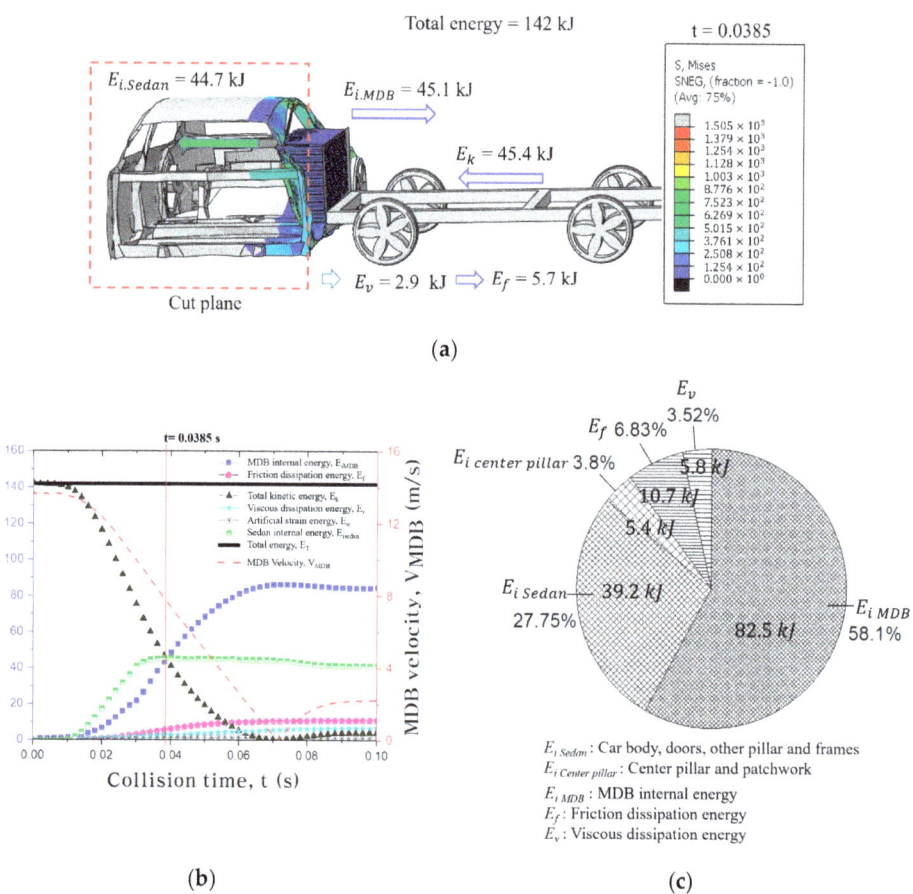

Figure 11. Energy conversion during side crash: (**a**) flow of energy distribution at t = 0.0385 s; (**b**) distribution of energies according to collision time; (**c**) distribution of energy conversion at t = 0.071 s.

Figure 11c shows the percentage of energy at t = 0.071 s, when MDB velocity was about 0 m/s. The $E_{i\cdot MDB}$ was 82.5 kJ, accounting 58.1% of the total energy, indicating that the 58.1% of total energy (142 kJ) was absorbed by the MDB during the side crash. On the other hand, the sum of $E_{i\cdot Sedan}$ absorbed by the doors, automobile body, and other frames was 44.6 kJ, with 5.4 kJ of energy absorbed by the center pillar and PW. According to the above data, the center pillar and PW took 12% of the sedan internal energy. The inner part of the center pillar was not included in this study. If the inner part of the center pillar been included, the center pillar, including the outer and inner portions, as well as the PW, would be expected to absorb 15~18% of the internal energy. The effect of PW and PS on energy distribution and intrusion resistance was investigated following, as reported below.

3.2.2. Effect of PW and PS on Elastic Strain Energy

Figure 12 shows the elastic strain energy (E_{es}) of the four kinds of center pillar (HPF, HPF + PW, HPF + PS, and HPF + PW + PS) during a side crash. For the HPF center pillar, the maximum E_{es} was 0.851 kJ at collision time (t) = 0.064 s and decreased to 0.30 kJ after the collision as a result of elastic recovery. When the PW was applied to the HPF center pillar, the maximum E_{es} increased to 0.928 kJ at t = 0.056 s. Because the rigidity of the HPF center pillar with PW increased, the absorption of elastic energy improved significantly. However, for the HPF center pillar with PS, the maximum E_{es} decreased to 0.655 kJ at t = 0.061 s. When the PS was applied to the bottom region of the HPF center pillar, because the stiffness of the HPF center pillar with PS decreased, the absorbed energy decreased in terms of elastic deformation.

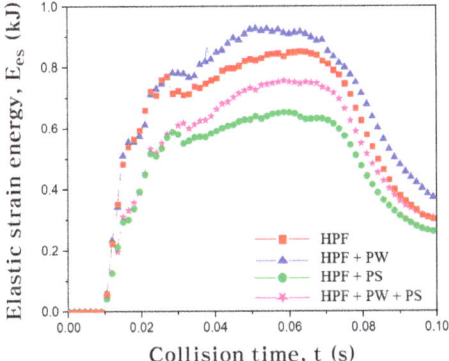

Figure 12. Elastic strain energy (E_{es}) of the four kinds of the center pillar during a side crash.

Figure 13 shows the maximum E_{es} of the four types of center pillar for detailed evaluation of energy absorption of PW and PS. The E_{es} of the HPF center pillar with PW was 0.1 kJ, accounting for 10% of the total energy absorbed by the HPF center pillar with PW. The E_{es} of the HPF center pillar with PS decreased to 0.655~0.660, regardless of the inclusion of PW.

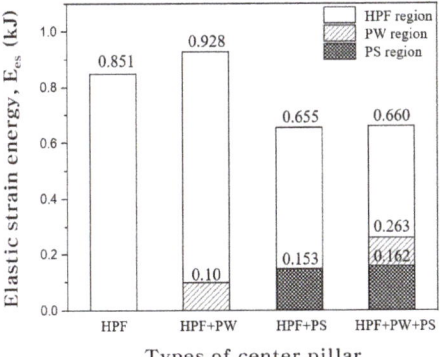

Figure 13. Maximum elastic strain energy (E_{es}) of the four types of center pillar.

3.2.3. Effect of PW and PS on Plastic Dissipated Energy

Figure 14 shows the plastic deformation energy (E_p) of the four kinds of the center pillar during a side crash. For the HPF center pillar, E_p increased to 4.538 kJ during deformation. The E_p was maintained after the collision, in contrast to the E_{es}, as a result of permanent deformation. For the HPF center pillar with PS, the E_p increased to 4.998 kJ at 0.1 s. The E_p of the HPF center pillar with PS was higher than that of the HPF center pillar. When PS was applied to the HPF center pillar, a large deformation occurred in the soft bottom region of the center pillar as a result of the ductility. As a result, the absorbed energy improved by 9.3%. However, there was no difference in absorbed energy between the HPF center pillar and the HPF center pillar with PW in terms of plastic deformation.

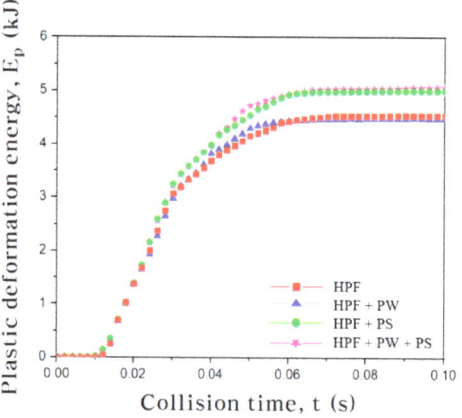

Figure 14. Plastic deformation energy (E_p) of the four kinds of the center pillar during side crash.

Figure 15 shows the maximum E_p of the four types of center pillar. For the HPF center pillar with PS and the HPF center pillar with PW and PS, 2.52~2.54 kJ of E_p was absorbed in the PS region, i.e., the bottom region of the center pillar. In terms of plastic deformation, the effect of PW was minimal, unlike elastic strain energy, whereas the E_p of the PS region accounted for 49.8~50.9% of the total plastic deformation energy between the HPF center pillar with PS and the HPF center pillar with PW and PS.

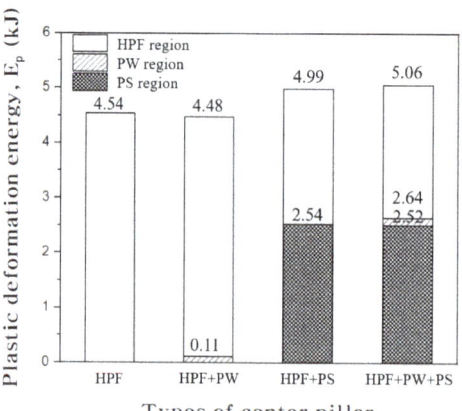

Figure 15. Maximum plastic deformation energy (E_p) of the four types of center pillar.

3.2.4. Effect of PW and PS on Internal Energy

Figure 16 shows the internal energy (E_i) of the four kinds of center pillar during a side crash. The total internal energy resulting from a collision consists of several energies and can be expressed by Equation (14):

$$E_i = E_{es} + E_p + E_a + E_{others} \qquad (14)$$

where E_i is the internal energy; E_{es} is the elastic strain energy; E_p is the plastic deformation energy; E_a is the artificial strain energy; and E_{others} is the energy dissipated by creep, viscoelasticity, and swelling.

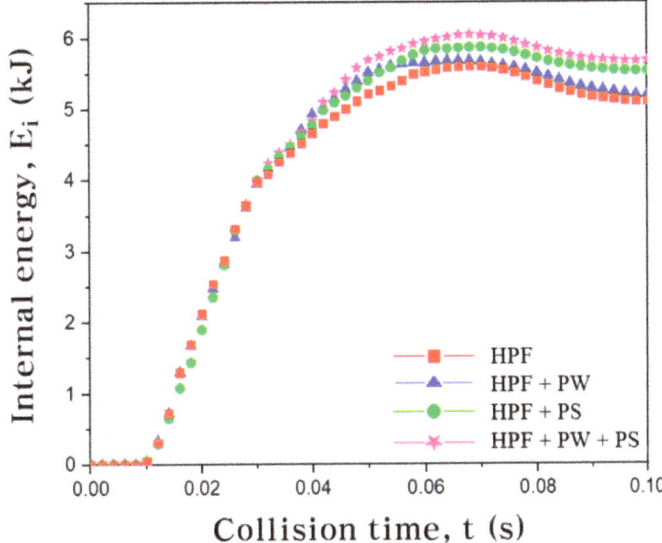

Figure 16. Internal energy (E_i) of the four kinds of center pillar during a side crash.

Generally, E_{es} and E_p are the dominant factors affecting E_i during a collision, and the contributions of other energies are relatively small. Therefore, in this study, the internal energy was expressed as the sum of E_{es} and E_p, excluding other types of energy, such as viscoelasticity, friction, and creep energy. In terms of internal energy, when the PS technique was applied to the HPF center pillar, there was a slight difference in the absorbed energy. However, a synergistic effect occurred when PW and PS were applied to the HPF center pillar, and a large amount of energy was absorbed, as shown in Figure 16.

Figure 17a shows the maximum E_i for the four types of center pillar. As shown in Figure 17a, there is almost no difference in the plastic deformation energy between the HPF center pillars with and without PW. The PW was welded on the upper part of the HPF center pillar to increase the stiffness, but the impact region was applied from the middle to the bottom, as shown in Figure 17b. Therefore, small deformation occurred in the PW. Likewise, a comparison between the HPF center pillar with PS and the HPF center pillar with PW and PS revealed similar phenomenon with respect to the absorbed plastic deformation energy. Based on the above results, it is necessary to review whether PS technology should applied in light of the relationship between technical performance and manufacturing cost.

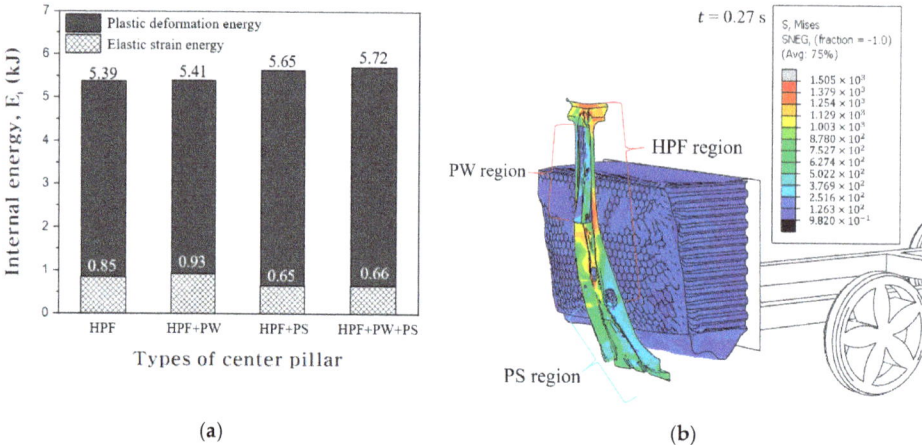

Figure 17. (a) Maximum internal energy (E_i) for the four types of center pillar. (b) Stress distribution of the center pillar during collision.

3.2.5. Effect of PW and PS on the Intrusion Resistance

During a side crash, a large amount energy is absorbed. However, if the material is soft and a large amount of deformation occurs, the intrusion displacement increases, putting the passenger in danger. Therefore, it is important to evaluate intrusion displacement for passenger safety, as well as collision energy absorption.

As shown in Figure 18a, because both the center pillar and the MDB were deformed during the collision, the maximum intrusion displacement was calculated according to Equation (15).

$$d_{max.i} = d_{max.DMB}(t) - L_b \tag{15}$$

where $d_{max \cdot DMB}$ is the maximum displacement of DBM, and L_b is the barrier length.

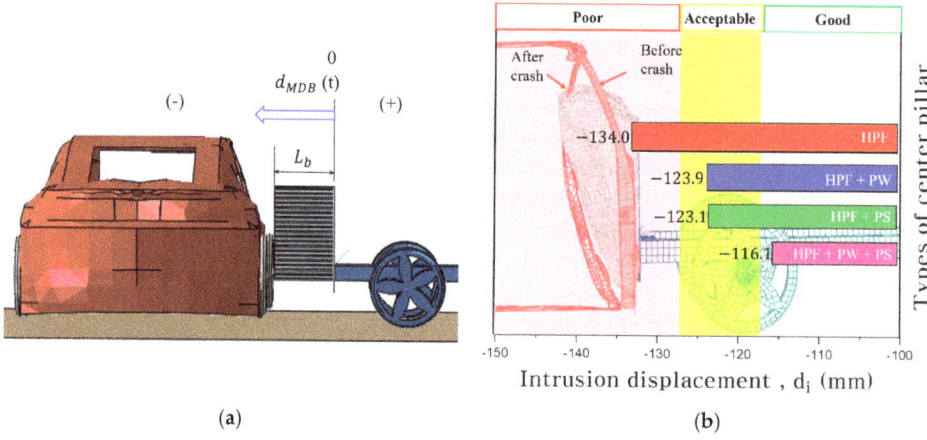

Figure 18. (a) Introduction of d_{MDB} and L_b. (b) Simulation data on intrusion displacement (d_i) of the four kinds of center pillar.

The evaluation method for the anti-intrusion resistance was introduced according to IIHS guidance. The primary performance of the center pillar is related to anti-intrusion resistance against side impact. According to the IIHS rating protocol, the center line of the vehicle's seat was generally in compliance with standard of the measured intrusion displacement, as shown in Figure 18b, with categories of good, acceptable, marginal, and poor [25]. In this study, multistructures, such as automobile seats, dummies, and windows were, not considered, so the maximum intrusion of the HPF center pillar with PW was used as the acceptable distance, achieving an acceptable level. As shown in Figure 18b, the maximum intrusion displacement of the HPF center pillar and the HPF center pillar with PW were −134.0 mm and −123.9 mm, respectively. When the stiffness of the upper part was strengthened with PW, the safety of the intrusion displacement was also improved by 7.5%. Likewise, in the case of the HPF center pillar with PS, the maximum intrusion displacement was −123.1 mm. A similar result was achieved with the HPF center pillar with PW. On the other hand, the maximum intrusion displacement of the HPF center pillar with PW and PS was −116.1 mm, representing an improvement of 13.4%. When the PW and PS were combined, a synergistic effect occurred.

4. Conclusions

In the side crash simulation performed in the present study, the effects of the HPF center pillar with PW and PS on energy distribution and anti-intrusion resistance were investigated, and the following conclusions were obtained:

1. During a side crash collision, the effect of PW in the HPF center pillar on plastic deformation energy was minimal but important with respect to elastic strain energy and intrusion resistance. When PW was applied to the HPF center pillar to strengthen the stiffness, the elastic strain energy and intrusion displacement were 10.0 and 7.5% respectively.
2. Because the PS technique provided ductility to the lower region of the HPF center pillar, large deformation occurred, and the plastic deformation energy was improved significantly by 10%. The plastic deformation energy of the PS region accounted for 49.8~50.9% of the plastic deformation energy of the HPF center pillar with PS.
3. In terms of total internal energy, the PS technique achieved better results than the PW technique. The maximum intrusion displacement of the HPF center pillar with PS was similar to that of the HPF center pillar with PW. In the case of the HPF center pillar with PW and PS, the maximum intrusion displacement was improved by 13.4%, and a synergistic effect of PW and PS occurred.

Author Contributions: Methodology, M.S.L.; writing—original draft, M.S.L.; writing—review and editing, M.S.L., C.K.J., and J.S.; software, M.S.L.; formal analysis, M.S.L.; investigation, O.D.L. and T.L.; funding, J.S.; validation and supervision: C.K.J. All authors have read and agreed to the published version of the manuscript.

Funding: This research was supported by the Ministry of Trade, Industry & Energy (20017450). This work was supported by the Energy technology development program (20206310200010, Advanced Remanufacturing of industrial machinery based on domestic CNC and building infrastructure for remanufacturing industry) funded By the Ministry of Trade, Industry & Energy (MOTIE, Korea).

Data Availability Statement: All datasets associated with this research are available upon request to the corresponding author.

Conflicts of Interest: The authors declare no conflict of interest.

References

1. Frómeta, D.; Lara, A.; Molas, S.; Casellas, D.; Rehrl, J.; Suppan, C.; Larour, P.; Calvo, J. On the correlation between fracture toughness and crash resistance of advanced high strength steels. *Eng. Fract. Mech.* **2019**, *205*, 319–332. [CrossRef]
2. Ma, B.L.; Wan, M.; Li, X.J.; Wu, X.D.; Diao, S.K. Evaluation of limit strain and temperature history in hot stamping of advanced high strength steels (AHSS). *Int. J. Mech. Sci.* **2017**, *128–129*, 607–613. [CrossRef]
3. Ashok, K.P.; Merbin, J.; Udaya, B.K.; Pradeep, L.M. Advanced High-Strength Steels for Automotive Applications: Arc and Laser Welding Process, Properties, and Challenges. *Metals* **2022**, *12*, 1051.
4. Christian, L.; Norbert, K.; Frank, B.K. Advanced High Strength Steels (AHSS) for Automotive Applications—Tailored Properties by Smart Microstructural Adjustments. *Steel Res. Int.* **2017**, *88*, 1700210.

5. Park, J.Y.; Jo, M.C.; Song, T.J.; Kim, H.S.; Sohn, S.S.; Lee, S.K. Ultra-high strength and excellent ductility in multi-layer steel sheet of austenitic hadfield and martensitic hot-press-forming steels. *Mater. Sci. Eng. A.* **2019**, *759*, 320–328. [CrossRef]
6. Kim, H.G.; Won, C.H.; Choi, S.G.; Gong, M.G.; Park, J.G.; Lee, H.J.; Yoo, J.H. Thermo-Mechanical Coupled Analysis of Hot Press Forming with 22MnB5 Steel. *Int. J. Automot. Technol.* **2019**, *20*, 813–825. [CrossRef]
7. Lee, M.S.; Moon, Y.H. Collision resistance of a lightweight center pillar made of AA7075-T6. *Int. J. Automot. Technol.* **2019**, *22*, 853–862. [CrossRef]
8. Liu, Y.; Zhu, Z.; Wang, Z.; Zhu, B.; Wang, Y.; Zhang, Y. Formability and lubrication of a B-pillar in hot stamping with 6061 and 7075 aluminum alloy sheets. *Procedia Eng.* **2017**, *207*, 723–728. [CrossRef]
9. Lee, M.S.; Seo, H.Y.; Kang, C.G. Comparative study on mechanical properties of CR340/CFRP composites through three point bending test by using theoretical and experimental methods. *Int. J. Precis. Eng. Manufac. Green Technol.* **2016**, *3*, 359–365. [CrossRef]
10. Çavuşoğlu, O.; Çavuşoğlu, O.; Yılmazoğlu, A.G.; Üzel, U.; Aydın, H.; Güral, A. Microstructural features and mechanical properties of 22MnB5 hot stamping steel in different heat treatment conditions. *J. Mater. Res. Tech.* **2020**, *9*, 10901–10908.
11. Omer, K.; Kortenaar, L.T.; Butcher, C.; Worswick, M.; Malcolm, S.; Detwiler, S. Testing of a hot stamped axial crush member with tailored properties—Experiments and models. *Int. J. Impact Eng.* **2017**, *103*, 12–28. [CrossRef]
12. Shi, D.Y.; Watanabe, K.; Naito, J.; Funada, K.; Yasui, K. Design optimization and application of hot-stamped B pillar with local patchwork blanks. *Thin Walled Struct.* **2022**, *170*, 108523. [CrossRef]
13. Mori, K.I.; Kaido, T.; Suzuki, Y.; Nakagawa, Y.; Abe, Y. Combined process of hot stamping and mechanical joining for producing ultra-high strength steel patchwork components. *J. Manuf. Process.* **2020**, *59*, 444–455. [CrossRef]
14. Ahmad, M.A.; Zakaria, A. Optimization of Spot Welds on Patchwork Blank for Hot Forming Process. *Appl. Mech. Mater.* **2014**, *606*, 177–180. [CrossRef]
15. Chengxi, L.; Zhongwen, X.; Weili, X. Hot stamping of patchwork blanks: Modelling and experimental investigation. *Int. J. Adv. Manuf. Technol.* **2017**, *92*, 2609–2617.
16. Gao, T.H.; Ying, L.; Dai, M.; Shen, G.; Hu, P.; Shen, L. A comparative study of temperature-dependent interfacial heat transfer coefficient prediction methods for 22MnB5 steel in spray quenching process. *Int. J. Therm. Sci.* **2019**, *139*, 36–60. [CrossRef]
17. Ota, E.; Yogo, Y.; Iwata, N. CAE-based process design for improving formability in hot stamping with partial cooling. *J. Mater. Process Technol.* **2019**, *263*, 198–206. [CrossRef]
18. Bok, H.H.; Choi, J.W.; Suh, D.W. Stress development and shape change during press-hardening process using phase-transformation-based finite element analysis. *Int. J. Plast.* **2015**, *73*, 142–170. [CrossRef]
19. Kim, D.K.; Woo, Y.Y.; Park, K.S. Advanced induction heating system for hot stamping. *Int J. Adv. Manufac. Technol.* **2018**, *99*, 583–593. [CrossRef]
20. Lee, M.S.; Lim, O.D.; Kang, C.G.; Moon, Y.H. Combined application of patchwork and partial softening to enhance the collision resistance of the center pillar. *Int. J. Crashworthiness* **2022**, *27*, 688–699. [CrossRef]
21. IMATEC. Available online: https://www.imatec.it/ (accessed on 11 May 2021).
22. Yin, H.; Xiao, Y.; Wen, G. Multiobjective Optimisation for Foam-Filled Multi-Cell Thin-Walled Structures Under Lateral Impact. *Thin Walled Struct.* **2015**, *94*, 1–12. [CrossRef]
23. Abaqus User's Manual 6.11. 2010. Available online: http://130.149.89.49:2080/v6.11/pdf_books/CAE.pdf (accessed on 11 May 2021).
24. Ciulia, V.; Sterfania, B.; Andrea, G.; Xiao, C. Numerical modeling of the 22MnB5 formability at high temperature. *Procedia Manuf.* **2019**, *29*, 428–434.
25. Young, J. *Side Impact Crashworthiness Evaluation; Crash Test Protocol Ver. IX User Guide*; Insurance Institute for Highway Safety (IIHS): Arlington, VA, USA, 2016.

Article

Evolution of Symmetrical Grain Boundaries under External Strain in Iron Investigated by Molecular Dynamics Method

Wenxue Ma [1], Yibin Dong [1], Miaosen Yu [1], Ziqiang Wang [1], Yong Liu [1], Ning Gao [1,2,*], Limin Dong [3] and Xuelin Wang [1,*]

[1] Institute of Frontier and Interdisciplinary Science and Key Laboratory of Particle Physics and Particle Irradiation (MOE), Shandong University, Qingdao 266237, China
[2] Institute of Modern Physics, Chinese Academy of Sciences, Lanzhou 730000, China
[3] Department of Materials Science and Chemical Engineering, Harbin University of Science and Technology, Harbin 150080, China
* Correspondence: ning.gao@sdu.edu.cn (N.G.); xuelinwang@sdu.edu.cn (X.W.)

Abstract: In the present work, the evolution of atomic structures and related changes in energy state, atomic displacement and free volume of symmetrical grain boundaries (GB) under the effects of external strain in body-centered cubic (bcc) iron are investigated by the molecular dynamics (MD) method. The results indicate that without external strain, full MD relaxations at high temperatures are necessary to obtain the lower energy states of GBs, especially for GBs that have lost the symmetrical feature near GB planes following MD relaxations. Under external strain, two mechanisms are explored for the failure of these GBs, including slip system activation, dislocation nucleation and dislocation network formation induced directly by either the external strain field or by phase transformation from the initial bcc to fcc structure under the effects of external strain. Detailed analysis shows that the change in free volume is related to local structure changes in these two mechanisms, and can also lead to increases in local stress concentration. These findings provide a new explanation for the failure of GBs in BCC iron systems.

Keywords: grain boundary; free volume; strain effect; micro-cracking; molecular dynamics

1. Introduction

It is well known that body-centered cubic (BCC) Fe-based steels have been extensively used in various industrial applications [1–3]. Typically, these steels are polycrystalline materials in which grain boundaries (GBs) are formed between crystallites [4]. Thus, the grain boundaries have a significant influence on the physical, mechanical and chemical properties of Fe-based materials. The investigation of their properties, including energy, structures and mechanics, presents a popular topic in the field of materials science [5,6]. For example, GB could act as a "sink" to absorb the interstitials and vacancies [7], which can decrease the number of radiation defects but can also affect the micro-structure of GB and related mechanical properties. Furthermore, external stress or strain fields, in addition to contribution from dislocations, GB sliding, migration, dislocation nucleation and failure are all involved in plastic deformation [8,9]. Since GB failure is a crucial factor for the design and application of Fe-based materials [10], it is necessary and important to study GB failure mechanisms at the atomic scale to explore the potential reasons for these phenomena.

The relationship between the atomic structure and energy of GBs has been studied extensively in the literature [11,12]. For example, Ratanaphan et al., calculated the energies of 408 grain boundaries in Fe and Mo using embedded atom method (EAM) potentials [5]. They reported that the calculated energies vary significantly with the grain boundary plane orientation but the energies do not show any distinct trends with misorientation angle or with the density of coincident lattice sites [5]. Gao et al. studied three typical GBs in BCC iron using molecular statics (MS) simulations, ab initio density-functional theory (DFT) calculations and the simulated high-resolution transmission electron microscopy (HRTEM) method, indicating the importance of relaxing the GBs in order to further investigate the properties of GBs through a multiscale method [13]. Du et al. also reported the energy states of some GBs in α-and γ-Fe using density-functional theory and found that the Σ3 twin boundaries exhibit low interface energies [14]. In addition to these results, the deformation behavior of GBs under different conditions is also a key factor to understand the GB properties. Spearot and colleagues investigated the influence of the structure of grain boundaries (e.g., bi-crystal GBs) on deformation behavior of the system by using MD simulations, through which the roles of dislocation nucleation and emission phenomenon under uniaxial tension were recognized for aluminum [15] and copper [16] symmetric tilt grain boundaries (STGBs). Terentyev et al. studied a set of <110> tilt grain boundaries (GB) in α-Fe with a misorientation angle varying from 26° to 141° by applying atomistic calculations and discovered grain boundary sliding is closely related to the structure determined by the misorientation angle [4]. Singh et al. reported the investigation of structure, energy and tensile behavior of niobium (Nb) bi-crystals containing symmetric and asymmetric tilt grain boundaries via molecular dynamics simulations [6]. Cui et al., further studied the <001>, <101> and <111> twist grain boundary structures in copper through molecular dynamics simulations to obtain the dependence of tensile strength on twist angle of grain boundary [17]. They reported that for a <001> and a <101> twist grain boundary, the tensile strength increases on average with an increasing misorientation twist angle [17]. However, the misorientation bears slight influence with respect to <111> twist GB structures [17]. From these results, it can be observed that various factors affect the structure and mechanical properties of GBs, highlighting the complicated nature of this topic within the material field. In order to solve these difficulties, researchers recently applied neutron diffraction method to investigate the free volume of sub-microcrystalline Ni [18], in which the anisotropic annealing of relaxed vacancies at GBs was identified, and is considered the main reason behind induction of the anisotropic length change upon annealing [18]. However, until now there are no detailed reports regarding the free volume change in GB under an external stress or strain field, thus warranting further investigated to better understand the properties of GBs.

As stated above, in addition to energy, structure and the deformation of grain boundary, the properties of free volume (FV) in grain boundaries should also be investigated in detail especially under the effect of an external stress or strain field. Generally, free volume in materials is generally defined as the maximum volume of a sphere that can be inserted between atomic sites in the system [19]. It has been reported that the free volume in grain boundary could assist the mobility of neighboring atoms, enabling GB sliding, grain rotation and GB dislocation emission [20]. By using molecular dynamics simulations through the investigation of three face-centered cubic (fcc) and body-centered cubic (bcc) metals, Sun et al. found that the free volume shrinks much faster above a critical temperature [21]. Tschopp et al. reported the dependence of free volume on the spatial correlation functions in grain boundary [20], which provides a better understanding of dislocation dissociation and nucleation in Cu grain boundaries. Furthermore, Tucker et al. investigated the effect of free volume on the stretching process of fcc copper grain boundaries and found that the free volume influences interfacial deformation through modified atomic-scale processes [22]. Wang et al. also suggested that FV may provide a site for micro-crack nucleation in bcc symmetrical grain boundaries after irradiation [23]. However, although research has been performed investigating the effect of free volume on the properties of grain boundaries, the relationship between free volume and grain boundary mechanical properties has not been reported in detail. For example, the relationship between the FV change and the stress concentration resulting in slip system activation and the role of FV change in GB failure, warrants further investigation. The results are expected to provide a comparison and validation with neutron diffraction results induced by the external stress or strain field in the future.

Therefore, in this paper, the properties of the symmetrical grain boundaries including $\Sigma 3(112)$, $\Sigma 3(111)$, $\Sigma 5(012)$, $\Sigma 5(013)$, $\Sigma 9(221)$, $\Sigma 11(113)$ and $\Sigma 17(410)$ in bcc iron are investigated at the atomic scale, and in particular, the influence of free volume on stress–strain properties. New mechanisms of GB failure by considering the free volume influence were explored in this work according to the simulation results. In the following, details of the simulation methods are introduced in Section 2, and the results and discussion are provided in Section 3. Finally, the conclusions are given in Section 4.

2. Method

In this work, molecular dynamics (MD) simulations are performed by Large scale Atomic/Molecular Massively Parallel Simulator (LAMMPS) [24] which is an open source MD software to simulate atomic interactions for a given system. The output data are analyzed by OVITO software (Version 3.5.4, OVITO GmbH, Darmstadt, Germany) [25]. Meanwhile, as stated above, 7 different grain boundaries including $\Sigma 3(112)$, $\Sigma 3(111)$, $\Sigma 5(012)$, $\Sigma 5(013)$, $\Sigma 9(221)$, $\Sigma 11(113)$ and $\Sigma 17(410)$ are built in bcc Fe based on coincidence site lattice (CSL) theory [26]. The GB angle of these GBs is from around 36° to 109°. Periodic boundary condition (PBC) is applied along 3 directions. In order to avoid the interaction between two GBs induced by PBC, the distance along the normal direction of GB is set to at least 14.8 nm, as listed in Table 1. A schematic of the symmetrical GB model is shown in Figure S1, provided in the Supplementary Materials. The other information for each simulation box is also included in Table 1. Atomic structures of these 7 GBs before full relaxation are shown in Figure 1. The Fe-Fe interaction is described by Fe potential developed by Mendelev et al., [27]. This potential has been well applied for grain boundary simulations [5,13,28] and is suitable for the present purpose.

Table 1. Parameters of symmetrical GBs used in the present work.

NO.	GB Plane (hkl)	Sigma (Σ)	Simulation Box Length(Å)	Number of Unit Cells along x, y, z and Normal Direction of GB Plane	Number of Atoms
1.	(112)	$\Sigma 3$	121.15 × 74.2 × 209.8	30 × 30 × 15, Z	162,000
2.	(111)	$\Sigma 3$	121.2 × 148.4 × 209.8	30 × 30 × 30, Y	324,000
3.	(310)	$\Sigma 5$	85.8 × 271.8 × 271.8	30 × 30 × 15, Z	270,000
4.	(210)	$\Sigma 5$	85.8 × 191.5 × 191.5	30 × 30 × 15, Z	540,000
5.	(221)	$\Sigma 9$	121.2 × 363.4 × 256.9	30 × 30 × 15, Z	972,000
6.	(113)	$\Sigma 11$	133.9 × 80.8 × 189.4	10×20×20, Z	176,000
7.	(410)	$\Sigma 17$	235.4 × 57.2 × 235.4	20 × 20 × 10, Z	272,000

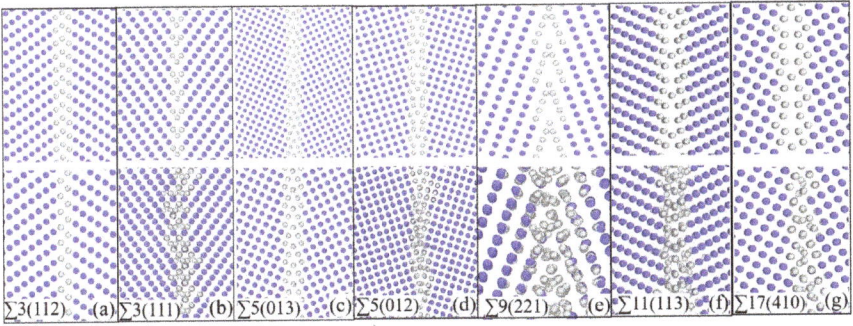

Figure 1. GB atomic structures before and after MD relaxations shown in the top and bottom panels, respectively: (**a**) $\Sigma 3(112)$ GB, (**b**) $\Sigma 3(111)$, (**c**) $\Sigma 5(013)$, (**d**) $\Sigma 5(012)$, (**e**) $\Sigma 9(221)$, (**f**) $\Sigma 11(113)$ and (**g**) $\Sigma 17(410)$. Structure types are analyzed by the common neighbor analysis method. Atoms near the GB region are colored white, while the atoms in the grain are colored blue.

Based on the above simulation models, conjugate-gradient (CG) method is used for relaxation at 0 K, and subsequently followed by MD relaxation at 300 K and 600 K. After full MD relaxation, the CG method is applied again to obtain the atomic structure and energy at 0 K. In this way, the local structure may overcome the energy barrier and reach a lower energy state after the CG-MD-CG relaxation process. In CG relaxation, the specified energy tolerance is 10^{-10} and the specified force tolerance is 10^{-10} eV/Å^3. During the MD relaxation, the timestep is 1 fs and at least 20 ps relaxation is applied for each process to ensure the system is fully relaxed at the given temperature. It should also be noted that during relaxation, the constant number of atoms, pressure and temperature (NPT) ensemble is applied in order to relax both the atomic position and simulation volume. Furthermore, for volume relaxation, each direction of the box is allowed to relax independently to fully release the internal elastic stress and obtain a more stable state for further calculations. The GB formation energy, E_{GB}, is then calculated according to the following equation as applied in previous studies [13,29,30], which is defined as the difference between the potential energy E_{total} of n atoms in the supercell containing GBs and the potential energy of a computational cell with the same number of atoms in a perfect crystal, divided by the cross-sectional area, S, of two GB planes.

$$E_{GB} = \frac{E_{total} - NE_p}{2S} \qquad (1)$$

where E_{total} is the total energy of a system containing 2 GBs. N is the number of atoms in this system. E_p is the cohesive energy of one atom in a perfect bcc Fe crystal, which is −4.12 eV according to Mendelev potential [27].

To investigate the mechanical property of the grain boundary, after full relaxation, external strain is applied to the system along the normal direction of GB plane with a strain rate of 5×10^9/s. The simulation temperature is set at 300 K. It should be noted that during the strain application along the normal direction of GB plane, the box length along the other two directions of the computational box is allowed to change automatically to ensure zero pressure along these two directions. The stress related to the increases in strain is also calculated and thus, the stress–strain curve is obtained. For comparison, this process is also performed for a perfect structure with strain direction along the same normal direction of GB plane. During the strain application along GB normal direction, the length of the box along the directions perpendicular to GB normal direction is also allowed to vary, which is similar to the real experimental condition and ensures the system is at a single stress–strain state [31].

To investigate the underlying factors affecting the mechanical property of GB, different methods are used to analyze GB structures and properties including the common neighbor analysis (CNA) [32], free volume (FV), single-atom potential energy distribution, local stress field and the single-atom displacement magnitude. In this work, CNA analysis, atomic energy distribution and displacement are performed by OVITO software and details can be found in [25]. The free volume is obtained by introducing enough grids along three directions in computation box and calculating the maximum distance (D_{max}) of each grid point to surrounding atoms. The free volume is then calculated as the sphere volume with radius of D_{max}-r, where r is the atomic radius in perfect lattice. This method has been well used in previous studies [33]. The FV calculation under the application of external stress was performed for polymer and amorphous materials [34], which have an FV distribution through the whole system. In this work, this method is also applied firstly for GB investigation since the FV in GB is expected to change in the GB failure process. Furthermore, as stated in the Introduction, the results from the present work may be used for comparison with neutron diffraction measurements under the effect of external stress field in future. The stress of each atom is calculated according to the following equation and can be viewed via the Ovito software:

$$\sigma_{ij}^V = \frac{1}{V}\sum_{\alpha}\left[\frac{1}{2}\sum_{\beta=1}^{N}\left(R_i^\beta - R_i^\alpha\right)F_j^{\alpha\beta} - m^\alpha v_i^\alpha v_j^\beta\right] \quad (2)$$

where (i,j) take values of x, y and z (directions). β takes values 1 to N neighbors of atom α. R_i^α is the position of atom α along direction i. $F_j^{\alpha\beta}$ is the force (along direction j) applied on atom β from atom α. V is the total volume of the system. m^α is the mass of atom α and v_i^α is the thermal excitation velocity of atom α along i direction.

3. Results and Discussion

3.1. Grain Boundary Energy and Structure after Different Relaxation Processes

Grain boundary energies of seven GBs calculated by using the method mentioned above are listed in Table 2, which includes the results following CG relaxation and CG-MD-CG relaxation. For comparison, the results from DFT calculations [35,36] are also listed. From this table, it is clear that CG-MD-CG relaxation results in lower energy state of GB according to Equation (1). The difference with and without MD relaxation is smaller for $\Sigma3(111)$, $\Sigma3(112)$ and $\Sigma5(013)$ GBs while it is larger for the other cases investigated in this work. The reason for this relates to the local atomic structure change induced by MD relaxation, as indicated by CNA and displacement results, shown in Figure 2. For example, $\Sigma3(112)$ and $\Sigma5(013)$ grain boundaries almost retain the symmetrical character with short atomic displacement distances after MD relaxation. However, for $\Sigma3(111)$ GB, after 600 K MD relaxation, some atoms located in the grain boundary area have moved a long distance, up to 1.3 Å, resulting in the loss of symmetry locally. In contrast, $\Sigma5(012)$, $\Sigma9(221)$, $\Sigma11(113)$ and $\Sigma17(410)$ GBs have almost lost the symmetrical character at the GB plane region as shown by displacement of the atoms in the GB region. The maximum atomic displacement of these cases is up to 4.81 Å. Therefore, the high temperature MD relaxations induced losses in symmetrical features in certain symmetrical GBs, through which these systems have overcome the energy barriers to lower energy states. In fact, a similar conclusion was made in previous studies [23,33], while in this work, more GBs confirm that the MD relaxation to a lower state is necessary for further analysis. The relationship between the GB formation energy and GB angle has also been investigated, as shown in Figure 2, which can be fitted by a Gaussian function (Equation (3)). Further, we have defined a new variable called $\Delta\sigma$, which is a fit for stress peak value by GB energy, as shown in Equation (4). $\Delta\sigma$ could help explain how the free volume had coupled with stress in the GB failure process.

$$E_{GB} = 1.228 - 0.958\exp\left[-0.5\left(\frac{\theta - 109.49}{5.1696}\right)^2\right] \quad (3)$$

Table 2. Grain boundary energy calculated by CG and CG-MD-CG relaxation processes. For comparison, the results from DFT calculations are also listed.

Sigma (Σ)	E_{GBs} J/m^2 (CG-MD-CG)	E_{GBs} J/m^2 (CG)	DFT
$\Sigma3(112)$	0.2703	0.3233	0.46 [35]
$\Sigma3(111)$	1.3144	1.3438	1.61 [36]
$\Sigma5(012)$	1.1301	1.6741	1.64 [35]
$\Sigma5(013)$	0.9889	1.0649	1.6 [35]
$\Sigma9(221)$	1.2639	2.3318	1.66 [37]
$\Sigma11(113)$	1.2551	2.4361	1.45 [37]
$\Sigma17(441)$	1.2356	2.5315	-

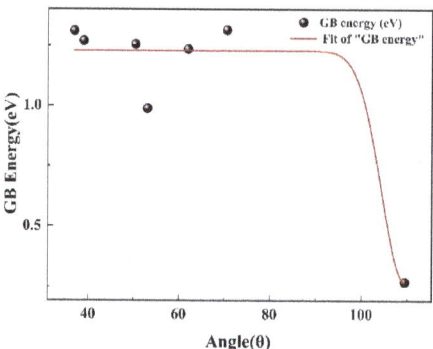

Figure 2. Gaussian Fit of the relationship between GB energy and GB misorientation.

3.2. Evolution and Related Failure Mechanisms of GBs under the External Strain Effect

In this section, the stress–strain curves for GBs and related single perfect crystal cases are calculated firstly, as shown in Figure 3. It is clear from Figure 3a, Σ3(111) and Σ17(410) GBs have the maximum and minimum peak tensile stress, respectively. In fact, a similar conclusion could also be made for single perfect crystal cases, as shown in Figure 3b. The peak tensile stresses of all cases studied in this work are listed in Table S1, in the Supplementary Materials. Compared with the results of single crystals, it is clear that the appearance of GBs induces the decrease in peak tensile stress.

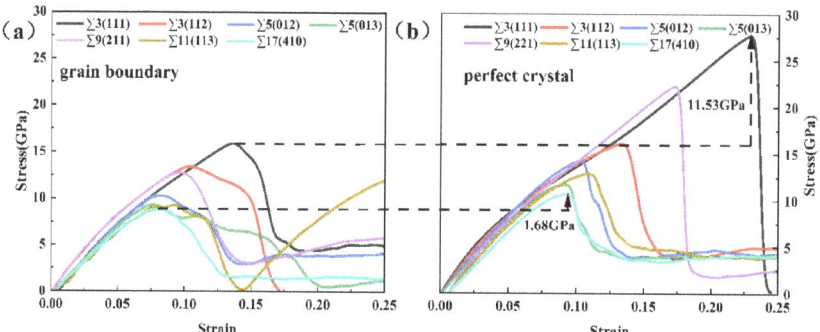

Figure 3. (a) Strain–stress curves of Σ3(111), Σ3(112), Σ5(012), Σ5(013), Σ9(221), Σ11(113), Σ17(410). (b) Strain–stress curves of single crystals corresponding to each grain boundary.

In order to explore the reason behind the peak value decrease induced by GB defect, the atomic stress state and related atomic structure were analyzed for these GB systems. Based on the analyses, the following mechanisms have been explored.

3.2.1. Phase Transformation Induced Grain Boundary Failure

Figure 4 depicts the structural evolution of the Σ3(112) and Σ5(013) GB system under the effect of external strain. From these results, it can be found that with increasing strain, the phase transition occurs initially in the GB region and extends to the grain interior, that is, from an initial bcc phase to fcc phase, as shown in Figure 4a. In the present work, this transition occurs with strain around 10% and related stress around 12.67 GPa, as shown in Figure 3a, and is also confirmed by the atomic potential energy and displacement, that is, atoms in the phase-transition region have higher potential energy and larger displacement distance, as shown in Figure S2 in the Supplementary Materials. In fact, the phase transition from bcc to fcc under external stress has also been reported in polycrystalline Fe systems with stress values up to around 13 GPa under shock wave [38]. It is clear that Σ3(112) GB has a similar effect on the phase transformation of Fe system. An example of a formed fcc structure is also shown in Figure S3 in the Supplementary Materials with a lattice constant around 3.7 Å, which is slightly larger than the value (3.6 Å) of fcc-Fe under the normal condition [39]. Together with this phase transition, a new interface is formed between the newly formed fcc phase and the original bcc phase, as shown in Figure 4b, whose position varies with the phase transition until the slip system {123}<111> becomes activated near the interface from the bcc phase side. Dislocation nucleation is then initiated from this activated slip system and finally the dislocation network is formed with increasing strain. One possible reason for the slip system activation in bcc instead of fcc phase for $\Sigma 3(112)$ GB can be explained by different Schmid factors (μ) in these two phases with external stress along <112> direction. In this case, the maximum Schmid Factor μ of bcc and fcc slip systems is 0.4115 and 0.4082, respectively. The larger Schmid factor in bcc phase indicates the it has higher probability to initiate the slip system in bcc phase. The value of 0.4115 in bcc phase is related to the {123}<111> slip system, same as the present simulation results, as shown in Figure 4b. In addition to above results, the larger displacement related to the phase transition is also confirmed to result in a larger free volume near the bcc-fcc interface, which is suspected as a possible reason for the change in stress field of the GB system. For example, comparing the maximum free volume in GB at states of 0 strain and 15% strain, as shown in Figure 5, it is clear that without strain, the maximum free volume (FV_{max}) is around 17.99 Å^3 and the average value of free volume is around 7.99 Å^3. While with 15% strain, FV_{max} reaches around 25.71 Å^3 with an average value around 11.98 Å^3 at the bcc-fcc interface induced by the phase transition. The stress field around this maximum free volume is around 21.48 GPa, resulting in the local stress concentration increasing and related activation of the slip system. Thus, the present results indicate that the increase in free volume induced by phase transition may be also one possible reason to induce the local stress to its critical value and the failure of GB system in bcc Fe.

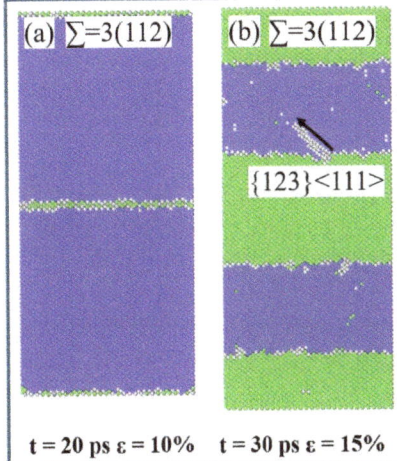
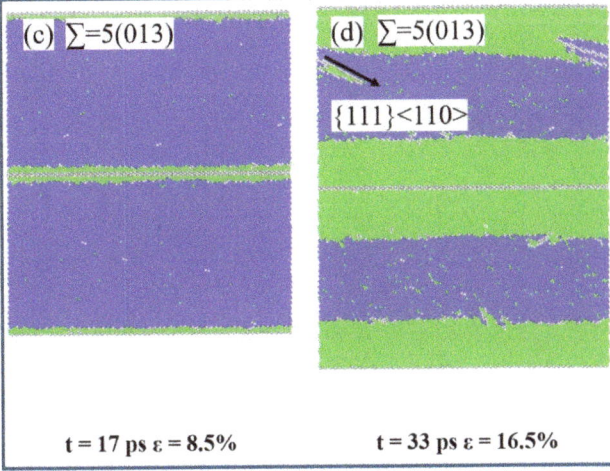

Figure 4. Snapshots of $\Sigma 3(112)$ and $\Sigma 5(013)$ GB under external strain effect. (**a**) Shows the state of $\Sigma 3(112)$ GB at which the phase transition starts with a strain value of 10% and (**b**) is the state at which the slip system in bcc phase near bcc-fcc interface is activated for $\Sigma 3(112)$. (**c**) Shows the state of $\Sigma 5(013)$ GB at which the phase transition starts with a strain value of 8.5% and (**d**) is the state at which the slip system is activated for $\Sigma 5(013)$ in fcc phase near the bcc-fcc interface. In the figure, the green, blue and white points are atoms in fcc, bcc and other phase states, respectively.

In fact, a similar process has also been observed in the $\Sigma 5(013)$ case, as shown in Figures S4 and S5 in the Supplementary Materials. The difference between these two cases is that in $\Sigma 5(013)$ case, there are still three atomic layers at the GB center without going through the phase transition process, as shown in Figure 4c. The phase transition occurs with strain up to 8%, at which the maximum free volume is also observed near the bcc-fcc interface with FV_{max} up to 24.43 Å^3. The local stress concentration is also observed above the maximum free volume region with a value around 21.65 GPa, resulting in the activation of slip system and related dislocation nucleation, as shown Figure 4d. Different to the activation of slip system initially in bcc phase for $\Sigma 3(112)$ GB, the activation of slip system under external stress along the <013> direction occurs initially in fcc phase in the {111} plane along <110> direction. Following the same method, the Schmid factor is also calculated for this case with external stress along <013> direction. The maximum Schmid factor for bcc and fcc phases are 0.4115 and 0.4899, respectively, in this case, which is the main reason for slip system activation near the interface from fcc phase side, as shown in Figure 4d.

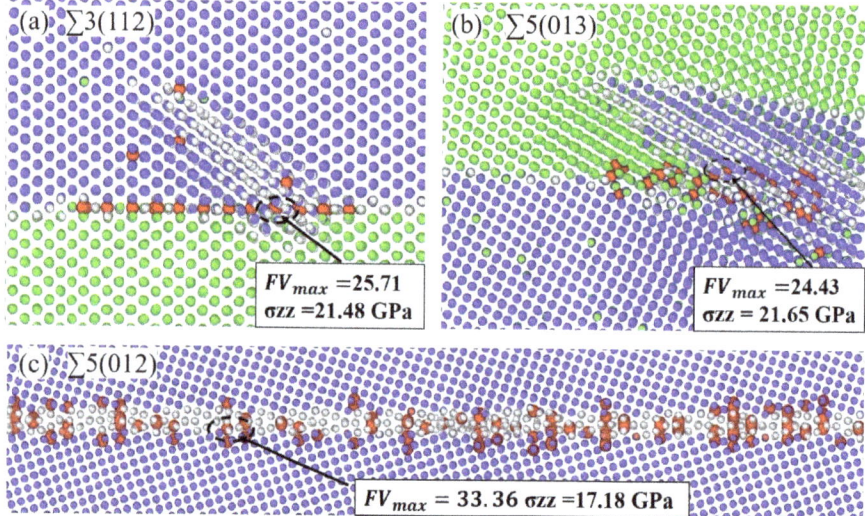

Figure 5. The maximum free volume and related stress distribution near the bcc-fcc interface when the stress is up to the peak value (shown by stress-strain curve in Figure 2) for (**a**) $\Sigma 3(112)$, (**b**) $\Sigma 5(013)$ and (**c**) $\Sigma 5(012)$ GB respectively. The blue and green balls are atoms in bcc and fcc state respectively. The larger red balls are free volume higher than 20 Å^3 in GB region.

3.2.2. Mechanical Failure Induced by Activation of Slip System from GB Plane

The second phenomenon accompanying the failure of GB explored in this work is the activation of slip systems directly at GB region without going through the phase transformation, as observed in the $\Sigma 5(012)$ and $\Sigma 3(111)$ $\Sigma 9(221)$, $\Sigma 11(113)$ and $\Sigma 17(410)$ GB cases. One example of $\Sigma 5(012)$ case is shown in Figure 6. As shown in Figure 6, the slip systems are activated from the GB plane to the grain interior at a time of around 16 ps after applying the external strain, at which the strain is around 8%. The Schmid factor is also calculated for this case. The results indicated the μ_{max} is up to 0.4625, which is related to the slip in {123} plane along <111> direction, as shown in Figure 6c. Furthermore, careful analysis of the local structure indicates local disordered regions in the GB region, which have high potential energy and high stress along the normal direction of GB plane. In fact, the local stress concentration is also observed in these regions with maximum stress around 27.4 GPa, as shown Figure S6 in the Supplementary Materials. Following the analysis method in Section 2, the atomic displacements were calculated around GB, indicating the maximum displacement distance also observed in these regions, as shown in the Supplementary Materials. The free volume change around these regions is then calculated. When the strain is 0, FV_{max} is around 31.24 Å^3, which is then increased to

33.36 Å³ above the disordered region, as shown in Figure 5c Therefore, the maximum free volume change is also one possible factor relating to the failure of Σ5(012) GB from the activation of slip system directly in local GB plane region. Further analysis of Σ3(111) Σ9(221), Σ11(113) and Σ17(410) GB reaches a similar conclusion. The example of Σ3(111) GB has been shown in Figures S7 and S8 in the Supplementary Materials. Based on these results, the derivative of critical stress along the normal direction of GB plane of GB failure, $\Delta\sigma$, can be described as a function of GB formation energy, E_{GB}, as shown in Figure 7a and the following equation:

$$\Delta\sigma = 1.241 - 1.297 exp\left[-0.5\left(\frac{E_{GB} - 4.073}{0.3886}\right)^2\right] \quad (4)$$

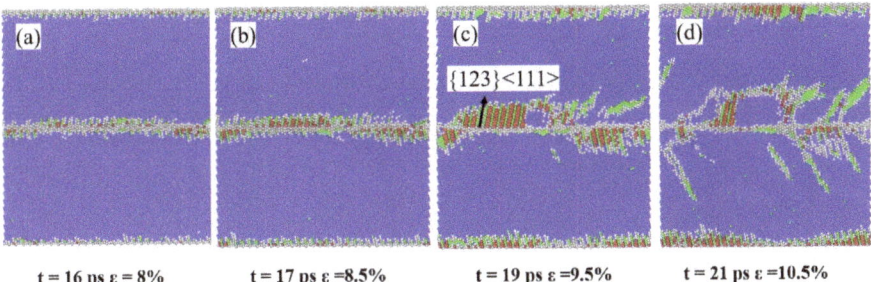

Figure 6. Snapshots of Σ5(012) GB evolution under external stress at different simulation times (t): (a) t = 16 ps, (b) t = 17 ps, (c) t = 19 ps and (d) t = 21 ps, respectively.

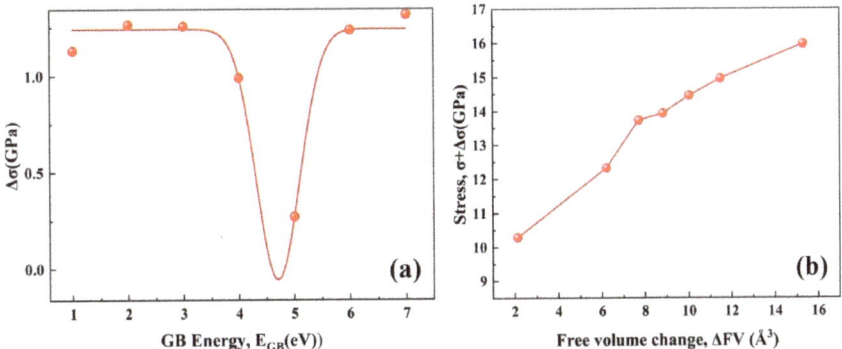

Figure 7. The Gaussian Fit of the relationship between GB energy and $\Delta\sigma$ (a). The relationship between free volume and fit stress. Fit stress equals the sum of the stress peak value and $\Delta\sigma$ (b).

Furthermore, the dependence of critical stress ($\sigma + \Delta\sigma$) of GB failure on free volume change (ΔFV) has also been explored, as shown in Figure 7b, which can be described by a new equation (Equation (5)). All of these results indicate that once the free volume change has been identified, the mechanical properties of GB can be estimated.

$$\sigma + \Delta\sigma = 8.21 \, \Delta FV^{0.2421} \quad (5)$$

4. Conclusions

In this work, the failure mechanism of symmetrical grain boundaries (GB) under external strain effects in body-centered cubic (bcc) iron are investigated via the molecular dynamics (MD) method. The local atomic structure evolution, energy state change, atomic displacements and free volumes were calculated for the above purpose. The following conclusions have been made:

(1) Full MD relaxations at high temperatures are necessary to obtain the lower energy states of GBs for further simulations under external strain.

(2) Two mechanisms are explored for the failure of symmetrical GBs under the external strain effect, including slip system activation, dislocation nucleation and dislocation network formation initially from the GB plane region induced by the external strain field or from the bcc-fcc phase interface induced by phase transformation under external strain effects.

(3) The change in free volume near the GB plane or bcc-fcc interface is not only related to the local structure change in the above two mechanisms, but can also lead to increases in the local stress concentration, providing a new explanation for the failure of GBs in BCC iron system.

Supplementary Materials: The following supporting information can be downloaded at: https://www.mdpi.com/article/10.3390/met12091448/s1, Figure S1: Schematic of grain boundary simulation model used in the present work. For stress-strain simulations, the external tensile strain field is applied on top and bottom surface of box, Figure S2: Atomic potential energy (**a**) and displacement distribution (**b**) of $\Sigma 3(112)$ GB at state with peak stress, Figure S3: Example of FCC lattice structure in fcc phase after phase transition, Figure S4: Snapshots (y-z plane) shows $\Sigma = 5(013)$ undergoes phase transition and green atom is fcc struc-ture blue atom is bcc. Then the GB happens to crack at 31 ps, region a, b and c are most obvious, Figure S5: Atomic potential energy (**a**) and displacement distribution (**b**) of $\Sigma 5(013)$ GB at state with peak stress, Figure S6: (**a**) The structure of $\Sigma 5(012)$ when slip system is activated with strain around 8%. The potential energy(**b**), stress(**c**) and atomic displacement distribution(**d**) at this state are shown respectively. Figure S7: (**a**) The structure of $\Sigma 3(111)$ when slip system is activated with strain around 14%. The potential energy, stress and atomic displacement distribution at this state are shown in (**b**), (**c**) and (**d**) respectively, Figure S8: Distribution of free volume near $\Sigma 3(111)$ GB region at (**a**) 0 ps and (**b**) at 28 ps (strain around 14%), Table S1: The peak tensile stresses of all cases studied in this work are listed in table.

Author Contributions: Conceptualization, N.G.; methodology, W.M. and N.G.; software, W.M.; validation, W.M.; formal analysis, W.M.; investigation, W.M.; resources, N.G.; data curation, W.M.; writing—original draft preparation, W.M.; writing—review and editing, Y.D., M.Y., Z.W., Y.L., N.G., L.D. and X.W.; visualization, W.M.; supervision, N.G.; project administration, N.G.; funding acquisition, N.G., X.W. and Y.L. All authors have read and agreed to the published version of the manuscript.

Funding: This research was funded by the National Natural Science Foundation of China, grant number (Project Nos. 12075141, 12175125 and 12105159).

Institutional Review Board Statement: Not applicable.

Informed Consent Statement: Not applicable.

Data Availability Statement: The data presented in this study are contained within the article.

Conflicts of Interest: The authors declare no conflict of interest.

References

1. Bonny, G.; Terentyev, D.; Elena, J.; Zinovev, A.; Minov, B.; Zhurkin, E.E. Assessment of Hardening Due to Dislocation Loops in Bcc Iron: Overview and Analysis of Atomistic Simulations for Edge Dislocations. *J. Nucl. Mater.* **2016**, *473*, 283–289. [CrossRef]
2. Messina, L.; Nastar, M.; Sandberg, N.; Olsson, P. Systematic Electronic-Structure Investigation of Substitutional Impurity Diffusion and Flux Coupling in Bcc Iron. *Phys. Rev. B* **2016**, *93*, 184302. [CrossRef]
3. Bauer, K.D.; Todorova, M.; Hingerl, K.; Neugebauer, J. A First Principles Investigation of Zinc Induced Embrittlement at Grain Boundaries in Bcc Iron. *Acta Mater.* **2015**, *90*, 69–76. [CrossRef]
4. Terentyev, D.; He, X.; Serra, A.; Kuriplach, J. Structure and Strength of <1 1 0> Tilt Grain Boundaries in Bcc Fe: An Atomistic Study. *Comput. Mater. Sci.* **2010**, *49*, 419–429. [CrossRef]
5. Ratanaphan, S.; Olmsted, D.L.; Bulatov, V.V.; Holm, E.A.; Rollett, A.D.; Rohrer, G.S. Grain Boundary Energies in Body-Centered Cubic Metals. *Acta Mater.* **2015**, *88*, 346–354. [CrossRef]

6. Singh, D.; Parashar, A. Effect of Symmetric and Asymmetric Tilt Grain Boundaries on the Tensile Behaviour of Bcc-Niobium. *Comput. Mater. Sci.* **2018**, *143*, 126–132. [CrossRef]
7. Samaras, M.; Derlet, P.M.; Van Swygenhoven, H.; Victoria, M. Radiation Damage near Grain Boundaries. *Philos. Mag.* **2003**, *83*, 3599–3607. [CrossRef]
8. Van Swygenhoven, H.; Derlet, P.M. Grain-Boundary Sliding in Nanocrystalline Fcc Metals. *Phys. Rev. B Condens. Matter Mater. Phys.* **2001**, *64*, 224105. [CrossRef]
9. Hasnaoui, A.; Van Swygenhoven, H.; Derlet, P.M. Cooperative Processes during Plastic Deformation in Nanocrystalline Fcc Metals: A Molecular Dynamics Simulation. *Phys. Rev. B Condens. Matter Mater. Phys.* **2002**, *66*, 184112. [CrossRef]
10. Chen, J.; Hahn, E.N.; Dongare, A.M.; Fensin, S.J. Understanding and Predicting Damage and Failure at Grain Boundaries in BCC Ta. *J. Appl. Phys.* **2019**, *126*, 165902. [CrossRef]
11. Gui-Jin, W.; Vitek, V. Relationships between Grain Boundary Structure and Energy. *Acta Metall.* **1986**, *34*, 951–960. [CrossRef]
12. Van Beers, P.R.M.; Kouznetsova, V.G.; Geers, M.G.D.; Tschopp, M.A.; McDowell, D.L. A Multiscale Model of Grain Boundary Structure and Energy: From Atomistics to a Continuum Description. *Acta Mater.* **2015**, *82*, 513–529. [CrossRef]
13. Gao, N.; Fu, C.C.; Samaras, M.; Schäublin, R.; Victoria, M.; Hoffelner, W. Multiscale Modelling of Bi-Crystal Grain Boundaries in Bcc Iron. *J. Nucl. Mater.* **2009**, *385*, 262–267. [CrossRef]
14. Du, Y.A.; Ismer, L.; Rogal, J.; Hickel, T.; Neugebauer, J.; Drautz, R. First-Principles Study on the Interaction of H Interstitials with Grain Boundaries in α- and γ-Fe. *Phys. Rev. B Condens. Matter Mater. Phys.* **2011**, *84*, 144121. [CrossRef]
15. Spearot, D.E.; Jacob, K.I.; McDowell, D.L. Nucleation of Dislocations from [0 0 1] Bicrystal Interfaces in Aluminum. *Acta Mater.* **2005**, *53*, 3579–3589. [CrossRef]
16. Spearot, D.E.; Tschopp, M.A.; Jacob, K.I.; McDowell, D.L. Tensile Strength of <1 0 0> and <1 1 0> Tilt Bicrystal Copper Interfaces. *Acta Mater.* **2007**, *55*, 705–714. [CrossRef]
17. Cui, C.; Yu, Q.; Wang, W.; Xu, W.; Chen, L. Molecular Dynamics Study on Tensile Strength of Twist Grain Boundary Structures under Uniaxial Tension in Copper. *Vacuum* **2021**, *184*, 109874. [CrossRef]
18. Kotzurek, J.A.; Hofmann, M.; Simic, S.; Pölt, P.; Hohenwarter, A.; Pippan, R.; Sprengel, W.; Würschum, R. Internal Stress and Defect-Related Free Volume in Submicrocrystalline Ni Studied by Neutron Diffraction and Difference Dilatometry. *Philos. Mag. Lett.* **2017**, *97*, 450–458. [CrossRef]
19. Farkas, D.; Van Petegem, S.; Derlet, P.M.; Van Swygenhoven, H. Dislocation Activity and Nano-Void Formation near Crack Tips in Nanocrystalline Ni. *Acta Mater.* **2005**, *53*, 3115–3123. [CrossRef]
20. Tschopp, M.A.; Tucker, G.J.; McDowell, D.L. Structure and Free Volume of <1 1 0> Symmetric Tilt Grain Boundaries with the E Structural Unit. *Acta Mater.* **2007**, *55*, 3959–3969. [CrossRef]
21. Sun, H.; Singh, C.V. Temperature Dependence of Grain Boundary Excess Free Volume. *Scr. Mater.* **2020**, *178*, 71–76. [CrossRef]
22. Tucker, G.J.; Tschopp, M.A.; McDowell, D.L. Evolution of Structure and Free Volume in Symmetric Tilt Grain Boundaries during Dislocation Nucleation. *Acta Mater.* **2010**, *58*, 6464–6473. [CrossRef]
23. Wang, X.Y.; Gao, N.; Setyawan, W.; Xu, B.; Liu, W.; Wang, Z.G. Effect of Irradiation on Mechanical Properties of Symmetrical Grain Boundaries Investigated by Atomic Simulations. *J. Nucl. Mater.* **2017**, *491*, 154–161. [CrossRef]
24. Thompson, A.P.; Aktulga, H.M.; Berger, R.; Bolintineanu, D.S.; Brown, W.M.; Crozier, P.S.; in 't Veld, P.J.; Kohlmeyer, A.; Moore, S.G.; Nguyen, T.D.; et al. LAMMPS—A Flexible Simulation Tool for Particle-Based Materials Modeling at the Atomic, Meso, and Continuum Scales. *Comput. Phys. Commun.* **2022**, *271*, 108171. [CrossRef]
25. Stukowski, A. Visualization and Analysis of Atomistic Simulation Data with OVITO–the Open Visualization Tool. *Model. Simul. Mater. Sci. Eng.* **2010**, *18*, 015012. [CrossRef]
26. Santoro, A.; Mighell, A.D. Coincidence-site Lattices. *Acta Crystallogr. Sect. A* **1973**, *29*, 169–175. [CrossRef]
27. Mendelev, M.I.; Han, S.; Srolovitz, D.J.; Ackland, G.J.; Sun, D.Y.; Asta, M. Development of New Interatomic Potentials Appropriate for Crystalline and Liquid Iron. *Philos. Mag.* **2003**, *83*, 3977–3994. [CrossRef]
28. Chaussidon, J.; Fivel, M.; Rodney, D. The Glide of Screw Dislocations in Bcc Fe: Atomistic Static and Dynamic Simulations. *Acta Mater.* **2006**, *54*, 3407–3416. [CrossRef]
29. Gao, N.; Ghoniem, A.; Gao, X.; Luo, P.; Wei, K.F.; Wang, Z.G. Molecular Dynamics Simulation of Cu Atoms Interaction with Symmetrical Grain Boundaries of BCC Fe. *J. Nucl. Mater.* **2014**, *444*, 200–205. [CrossRef]
30. Wang, X.Y.; Gao, N.; Xu, B.; Wang, Y.N.; Shu, G.G.; Li, C.L.; Liu, W. Effect of Irradiation and Irradiation Defects on the Mobility of Σ5 Symmetric Tilt Grain Boundaries in Iron: An Atomistic Study. *J. Nucl. Mater.* **2018**, *510*, 568–574. [CrossRef]
31. Gao, N.; Setyawan, W.; Kurtz, R.J.; Wang, Z. Effects of Applied Strain on Nanoscale Self-Interstitial Cluster Formation in BCC Iron. *J. Nucl. Mater.* **2017**, *493*, 62–68. [CrossRef]
32. Faken, D.; Jónsson, H. Systematic Analysis of Local Atomic Structure Combined with 3D Computer Graphics. *Comput. Mater. Sci.* **1994**, *2*, 279–286. [CrossRef]
33. Gao, N.; Perez, D.; Lu, G.H.; Wang, Z.G. Molecular Dynamics Study of the Interaction between Nanoscale Interstitial Dislocation Loops and Grain Boundaries in BCC Iron. *J. Nucl. Mater.* **2018**, *498*, 378–386. [CrossRef]
34. Hossain, D.; Tschopp, M.A.; Ward, D.K.; Bouvard, J.L.; Wang, P.; Horstemeyer, M.F. Molecular Dynamics Simulations of Deformation Mechanisms of Amorphous Polyethylene. *Polymer* **2010**, *51*, 6071–6083. [CrossRef]
35. Wang, J.; Janisch, R.; Madsen, G.K.H.; Drautz, R. First-Principles Study of Carbon Segregation in Bcc Iron Symmetrical Tilt Grain Boundaries. *Acta Mater.* **2016**, *115*, 259–268. [CrossRef]

36. Bhattacharya, S.K.; Tanaka, S.; Shiihara, Y.; Kohyama, M. Ab Initio Study of Symmetrical Tilt Grain Boundaries in Bcc Fe: Structural Units, Magnetic Moments, Interfacial Bonding, Local Energy and Local Stress. *J. Phys. Condens. Matter* **2013**, *25*, 135004. [CrossRef]
37. Wang, J.; Madsen, G.K.H.; Drautz, R. Grain Boundaries in Bcc-Fe: A Density-Functional Theory and Tight-Binding Study Grain Boundaries in Bcc-Fe: A Density- Functional Theory and Tight-Binding Study. *Model. Simul. Mater. Sci. Eng.* **2018**, *26*, 025008. [CrossRef]
38. Gunkelmann, N.; Bringa, E.M.; Kang, K.; Ackland, G.J.; Ruestes, C.J.; Urbassek, H.M. Polycrystalline Iron under Compression: Plasticity and Phase Transitions. *Phys. Rev. B Condens. Matter Mater. Phys.* **2012**, *86*, 144111. [CrossRef]
39. Pan, F.; Zhang, Z.S. Magnetic Property of the Face Center Cubic Iron with Different Lattice Parameter in Fe/Pd Multilayers. *Phys. B Condens. Matter* **2001**, *293*, 237–243. [CrossRef]

Article

The Role of Grain Boundaries in the Corrosion Process of Fe Surface: Insights from ReaxFF Molecular Dynamic Simulations

Zigen Xiao [1], Yun Huang [2], Zhixiao Liu [1,*], Wangyu Hu [1], Qingtian Wang [3] and Chaowei Hu [3,*]

1. College of Materials Science and Engineering, Hunan University, Changsha 410082, China; xiaozigeng@hnu.edu.cn (Z.X.); wyuhu@hnu.edu.cn (W.H.)
2. School of Physics and Electronics, Hunan University, Changsha 410082, China; huangyun19@hnu.edu.cn
3. Key Laboratory of Nuclear Reactor System Design Technology, Nuclear Power Institute of China, Chengdu 610041, China; wqtian@126.com
* Correspondence: zxliu@hnu.edu.cn (Z.L.); npichd10@npic.ac.cn(C.H.)

Abstract: Intergranular corrosion is the most common corrosion phenomenon in Fe-based alloys. To better understand the mechanism of intergranular corrosion, the influence of grain boundaries on Fe-H_2O interfacial corrosion was studied using molecular dynamics simulation based on a new Fe-H_2O reaction force field potential. It is found that the corrosion rate at the polycrystalline grain boundary is significantly faster than that of twin crystals and single crystals. By the analysis of stress, it can be found that the stress at the polycrystalline grain boundary and the sigma5 twin grain boundary decreases sharply during the corrosion process. We believe that the extreme stress released at the grain boundary will promote the dissolution of Fe atoms. The formation of vacancies on the Fe matrix surface will accelerate the diffusion of oxygen atoms. This leads to the occurrence of intergranular corrosion.

Keywords: ReaxFF; corrosion; Fe-H_2O interface; grain boundary

1. Introduction

Fe-based alloys are widely used as structural materials [1] because of their excellent tensile strength and high toughness [2]. However, the corrosion of metal in the working environment will reduce the mechanical properties of metal and shorten its service life. In the corrosion process, the metal is electrochemically oxidized into ions or some compounds [3]. In particular, the presence of grain boundaries can cause intergranular corrosion. The dissolution rate of Fe atoms in the grain boundary region is much greater than the dissolution rate of crystal grains, leading to local corrosion [4]. Intergranular corrosion refers to the boundaries of crystallites of the material being more susceptible to corrosion than their insides, which is mainly caused by the difference in chemical composition between the surface and the interior of the grains and the existence of internal stress.

The surface structure of the Fe-based alloys is complex with defects such as grain boundaries, corners, edges, boundaries, interference layers, etc. [5,6]. Corrosion usually occurs at the defect first, and the corrosion at the defect is more severe than inside the crystal. It can be observed in the experiments that the corrosion is more severe at the grain boundary compared with the inside of the crystal [7]. You et al. found that the corrosion in pure Fe is initiated by pits that quickly become blocked and that it propagates around the pits giving rise to the formation of porous layers [8]. The grain refinement obtained by rolling improved the corrosion resistance of iron in sulfuric acid solution, borate buffer solution, and borate buffer solution with chloride ion [9]. Bennett et al. studied the influence of orientation angle and grain boundary structure on grain boundary corrosion of austenitic and ferritic stainless steels. They pointed out that the grain boundary corrosion mechanism is not only the dissolution of the chromium-depleted alloy but also other mechanisms, but the specific mechanism is not distinct [10]. According to Lapeire's study, grain orientation can also affect corrosion behavior, and the orientation of adjacent grains plays a dominant role in the dissolution rate [11]. The reduction of the electronic work function at the grain boundary indicates the electrons at the grain boundary are more active, which makes the grain boundary vulnerable to electrochemical attack [12]. Emily et al. believed that the corrosion at the grain boundary is caused by the combined effect of sensitization and crevice corrosion. The precipitation of chromium and the grain boundary has set the local corrosion sensitivity [13].

Up to now, the density functional theory calculations (DFT) and reaction force field (ReaxFF) molecular dynamics (MD) simulations have been usually used to study Fe and Fe-water surface corrosion mechanisms. Hu et al. studied the effects of four typical surface adsorbates [14]. Their works suggested that the strong interaction between oxygen and surface will weaken the bonds' strength between substrate atoms, which will lead to structure deformation and charge redistribution in the substrate. In addition, Bucun et al. studied the effects of oxygen precovering and water adsorption on the surface of Fe substrates [15]. The time and space scales of the above simulations are seriously limited, which can only describe the trend and cannot carry out the complete dynamic evolution of the corrosion process. The ReaxFF-MD provides an alternative method to simulate the interface reaction and can explain the effect of defects (such as grain boundaries) on corrosion from stress changes and energy changes [16,17]. Verners et al. [18] studied nickel-based alloys' stress corrosion cracking behavior. The existence of stress will hinder the twin dislocation failure and activate the new slip to reduce the strength of nickel. DorMohammadi et al. used ReaxFF-MD to study the initial stage of Fe corrosion in pure water under different applied electric fields and temperatures and then studied the passivation mechanism of Fe substrate and the depassivation process of chloride [19–21]. The study of Klu et al. showed that the existence of polycrystalline grain boundaries could significantly increase the diffusion coefficient of carbon, and the release of stress can reduce the diffusion barrier of carbon [22].

In summary, defects in pure iron will preferentially corrode during the corrosion process. Whether the grain boundary as a typical defect will affect the surface corrosion process of pure iron. The computational methods involving simulations of dynamic processes are an important tool for understanding the mechanism of dynamic processes in reactive systems, such as corrosion. The present article, using a reactive force field (ReaxFF) molecular dynamics (MD) simulation, highlights the effect of grain boundaries on the evolution of the corrosion process in a reactive pure Fe-H_2O system.

2. Computational Details

In this study, three kinds of Fe models—single crystal, sigma3 or sigma5 twins, and polycrystalline—were used to study the influence of grain boundary on the corrosion process of pure Fe-H_2O surface systems. The sigma3 twin boundary is composed of the (111) plane, and the sigma5 twins plane is composed of the (02-1) plane. The structure at the twin grain boundary is stable and the interfacial energy is low. However, the polycrystalline grain boundary structure is unstable, the lattice distortion is large, and there are many defects. The stress in this paper is the shear stress of the Fe substrate on the Y-axis. It is obtained by counting the stress sum of Fe atoms at the grain boundary in the Y direction and dividing it by the corresponding total atomic volume. The volume of polycrystalline grains is 35 nm^3. The schematic diagram of the models is shown in Figure 1. The Fe substrate comprises twelve atom layers in the polycrystalline simulation system, including 9215 Fe atoms and existing many defects such as grain boundaries and stacking fault. The water-side part includes the water layers of 40 Å thickness with the density of 0.99 g/cm^3 (about 4669 water molecules at 300 K and 1 atm).

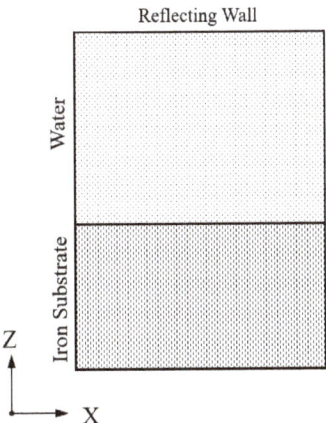

Figure 1. Schematic diagram of the simulation model for the Fe-water surface system.

All the simulations were based on the open-source Large-scale Atomic/Molecular Massively Parallel Simulator (LAMMPS) code [23] and the atomic interactions were described using the ReaxFF for pure Fe-H_2O systems. The original ReaxFF was developed by van Duin et al. [24]. The ReaxFF-MD can simulate the chemical reaction process that includes chemical bonds breaking and forming between atoms; therefore, it is suitable for studying interfacial corrosion. Recently, our group developed a new Fe-H_2O ReaxFF potential function [25], which can accurately describe the nature of the defects in Fe substrate and the Fe-water interactions. The interaction between vacancy and hydrogen/oxygen is emphasized to describe the precipitation of hydrogen/oxygen near the vacancy. The energy contributions to the Fe-H_2O ReaxFF potential are summarized by the following [25]:

$$E_{system} = E_{bond} + E_{over} + E_{under} + E_{val} + E_{lp} + E_{Coul} + E_{H-bond} + E_{vdW} \quad (1)$$

which includes terms related to bond, angle, over coordination, undercoordination, lone pair, van der Waals, Coulomb, and H-bond energies. The details of the ReaxFF potential have been shown in our recent paper. The essential part of ReaxFF is that the charge equilibration (QEq) method can be used to obtain the charge distribution [26,27]. The charge values were determined at each simulation time step and depended on the system's geometry. This feature made it possible to describe charge transfer in chemical reactions using ReaxFF. Subsequently, this study did not consider an absolute stress value analysis, instead focusing on relations between reactivity and stress states.

The periodic boundary conditions were applied in the X- and Y-directions in all simulated boxes, and the Z-direction uses aperiodic boundary conditions. A reflective wall is added at the upper Z boundary to prevent atoms from passing through the boundary. The Fe substrate and water molecules are relaxed with a Nose/Hoover isothermal–isobaric (NPT) [28] to reach equilibrium. Nose–Hoover thermostat [29,30] is used to maintain the prescribed system temperature during corrosion for the canonical (NVT) ensemble. The MD simulation time step is 0.2 fs. The corrosion process of the pure Fe-H_2O system was observed by external electric field [31] acceleration at 300 K and NVT ensemble, which is located at 0~3 Å on the surface of the Fe substrate. We chose the most suitable external electric field to be 325 MV/cm by observing the corrosion phenomenon. All subsequent simulations are performed under this electric field strength. The relationship between individual atomic charges and the corresponding electrostatic energy $E(\mathbf{q})$ with time is shown in Equation (2):

$$E(\mathbf{q}) = \sum_{i=1}^{N}\left[\chi_i q_i + \eta_i q_i^2 + Tap(r_{ij})k_c \frac{q_i q_j}{\left(r_{ij}^3 + \gamma_{ij}^{-3}\right)}\right] \quad (2)$$

where $\mathbf{q}$ represents a vector of length N containing the charges, q_i is the charge of ion i, N is the total number of ions, k_c is the dielectric constant, χ_i and η_i are represent the electronegativity and the hardness of ion i, $Tap(r)$ is a seventh-order taper function, and γ_{ij} refers to the shielding parameter between the two atoms I and j.

We reproduced their works to verify the correctness of the ReaxFF potential according to the article of DorMohammadi et al. [19]. The simulation models mainly consist of body-centered cubic (bcc) Fe and water molecules. The corrosion processes of low-index surfaces (100), (110), and (111) were accelerated by applying an external electrical field. The corrosion phenomenon is consistent compared with the works of DorMohammadi et al. The difference is that the diffusion of H atoms into the substrate can be clearly observed during the corrosion process. Therefore, we believe that the ReaxFF potential parameters can simulate Fe-water interfacial corrosion accelerated by an external electric field at room temperature. The relevant verification process is in the supporting materials.

3. Results

The Corrosion Process at the Grain Boundary

In order to study the corrosion process of a perfect crystal, we simulate the pure Fe-H_2O interfacial corrosion of Fe (110) single crystal for that the (110) surface is the most stable close-packed one. The corrosion processes of the single crystal are shown in Figure 2. It can be found that water molecules are adsorbed on the surface of the Fe substrate at 0.14 ps (Figure 2a); then the hydrogen-oxygen bond of the water molecule begins to stretch (Figure 2b); and finally, the water molecules decomposed completely into OH^- and free H^+ ions (Figure 2c). The chemical reaction during the corrosion process is shown below:

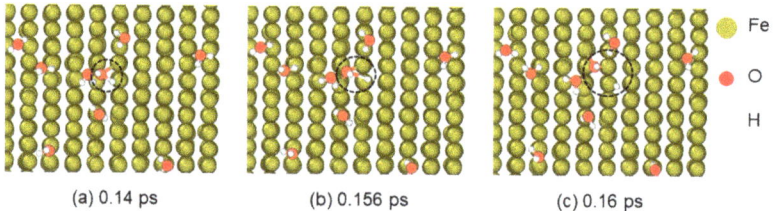

(a) 0.14 ps   (b) 0.156 ps   (c) 0.16 ps

Figure 2. (a) The adsorption of water molecules on the Fe substrate; (b) the elongation of the O-H bond in the adsorbed water molecules; and (c) the breaking of the O-H bond and the formation of OH^-.

Step 1: The water molecules adsorbed on the surface of the Fe substate dissociate to form OH^- and H^+:

$$H_2O \rightarrow OH^- + H^+$$

OH^- is adsorbed on the surface of the substrate to generate Fe(OH):

$$Fe + OH^- \rightarrow FeOH + e^-$$

The charge of the Fe atoms on the surface rises to 0.4 e, and the charge of the O atoms drops to -0.69 e.

Step 2: The Fe(OH) on the surface dissolves into the water:

$$Fe + FeOH + OH^- \rightarrow FeOH + FeOH^+ + 2e^-$$

The charge of the Fe atoms on the surface rises to 0.57 e

Step 3: Fe is oxidized to form Fe^{2+}:

$$FeOH^+ + H^+ \rightarrow Fe^{2+} + H_2O$$

Step 4: The adsorbed OH^- also dissociate into O and H ions, forming additional H^3O^+ and iron oxides.

During the process of the adsorption of H_2O on the surface of Fe substrate to form $Fe(OH)_2$, the surface Fe atoms are oxidized, and the average charge increases to 0.7 e. The O atoms adsorbed on the surface are reduced and the average charge drops to −0.775 e, H ions also penetrate into the Fe substrate with an average charge of −0.25 e. Finally, the adsorbed OH^- also dissociate into O and H ions, forming additional H_3O^+ and iron oxides.

In Fe single crystal, O ions penetration uniformly into the Fe substrate to form oxides (Figure 3a). The Fe (110) surface with a sigma3 twins is also studied, as shown in Figure 3b. It can be found that the corrosion process of the sample with sigma3 twins is similar to that of single crystals. After water molecules decompose on the Fe surface, Fe atoms will dissolve into the water solution, O ions will penetrate into the Fe substrate to form oxides, and H ions will diffuse into the Fe substrate. The existence of the twin grain boundary does not affect the corrosion rate of the Fe substrate. This situation is because the sigma3 grain boundaries are relatively stable and not easy to corrode.

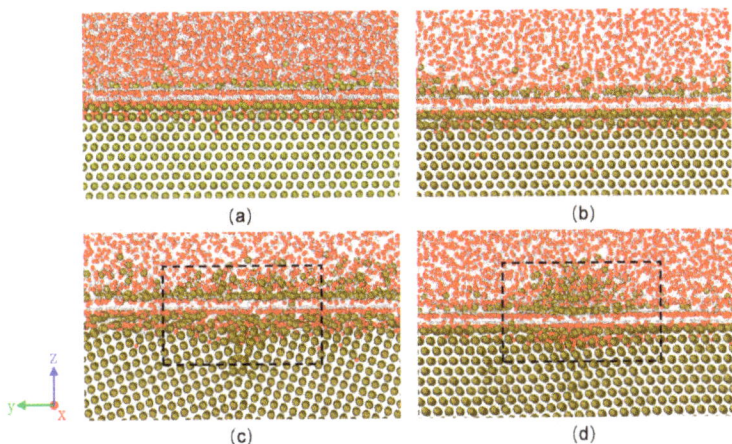

Figure 3. The snapshots at different corrosion stages for the different pure Fe-H_2O models in 300 K: (**a**) Fe single crystal (110); (**b**) Fe sigma3 (110) [111] twins; (**c**) Fe sigma5 (02-1) [012] twins; (**d**) polycrystals. The yellow, red, and white spheres represent Fe, oxygen, and hydrogen atoms.

However, the corrosion of sigma5 twins is different from that of sigma3 twins. The most stable (210) surface is used to test the effect of different grain boundaries on corrosion. As shown in Figure 3c, the penetration of O ions at the grain boundary is faster, and the corrosion depth at the grain boundary is more than four atomic layers. This phenomenon is related to the structure and stability of the grain boundary. As the angle of the grain boundary increases, the corrosion at the grain boundary will be easier.

To verify the above speculation, we simulate the corrosion of the polycrystalline grain boundary. According to Figure 3d, polycrystalline consists of three grains, and irregular grain boundaries are obtained by rotating the grain in the middle. The corrosion phenomenon of polycrystal is similar to sigma5 twins, but the phenomenon of grain boundary corrosion is more prominent. Many Fe atoms on the grain boundary are dissolved and O ions penetrate rapidly into the grain boundaries to form oxides. The number of dissolved Fe atoms in polycrystals is far greater than that of sigma5 twins, and the corrosion rate of grain boundaries is speedy. This may be related to the reduction of Fe work function (ionization energy) during H_2O adsorption obtained by the multielectron intercalation theory [17] and may also be related to stress.

It is found that O ions penetrate faster at the polycrystalline and sigma5 twin grain boundaries, and local corrosion occurs at the grain boundaries, as shown in Figure 4. In the polycrystalline model, O ions first penetrate into the grain boundaries and then along the grain boundaries to the surroundings to form oxides. In the twinned system, a small number of O ions are distributed on the sigma5 twin grain boundary. However, it is impossible to judge whether it penetrates along the grain boundaries, and the O ions in the sigma3 twins exhibit uniform penetration. During the corrosion process, the stress changes at the grain boundaries, as shown in Figure 5. The stress at the polycrystalline grain boundary decreases in a wide range during the corrosion process, and the stress at the sigma5 twin is only reduced at the grain boundary, but the stress at the sigma3 twin remains unchanged. The variation range of stress is consistent with the penetration range of O ions.

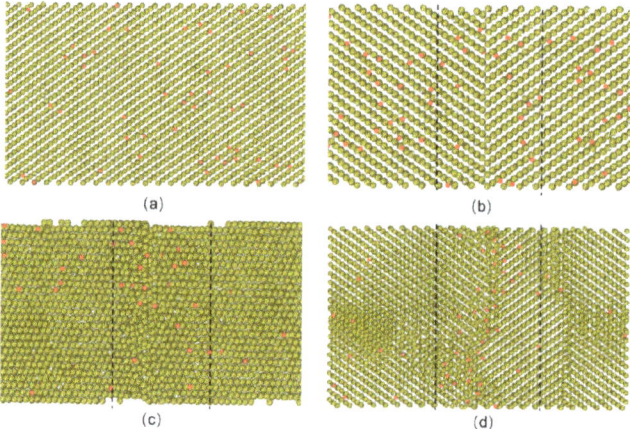

Figure 4. Distribution of oxygen atoms in Fe substrate during corrosion. (**a**) Fe single crystal (110); (**b**) Fe sigma3 (110) [111] twins; (**c**) Fe sigma5 (02-1) [012] twins; (**d**) polycrystals.

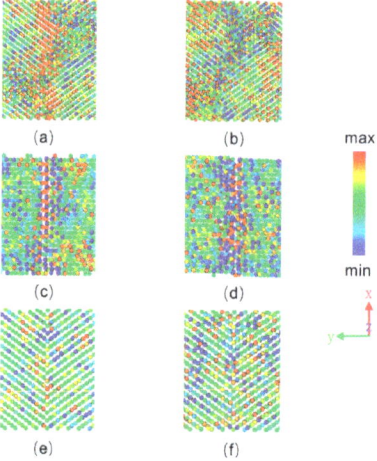

Figure 5. The stress change of the Fe substrate during the corrosion process. (**a,b**) are the polycrystalline; (**c,d**) are the sigma5 twins; (**e,f**) are the sigma3 twins.

4. Discussion
4.1. The Principle of Grain Boundary Corrosion

Figure 6 shows the variation of shear stress in the Y-direction of the iron substrate; the vertical axis is the difference between the initial stress and the instantaneous stress. During the corrosion process, the stress at the polycrystalline grain boundary was reduced by 2.0 GPa within 200 ps, the stress at the sigma5 twin boundary was reduced by 1.25 GPa, and the sigma3 twin boundary was only reduced by 0.75 GPa, but the stress of the single crystal remained unchanged. The changing trend of stress corresponds to the corrosion phenomenon of the model. Therefore, we believe that the enormous stress relief in the high-stress region at the grain boundary will deform the grain boundary structure and cause the atoms at the grain boundary to dissolve, leaving many vacancies on the surface. The existence of vacancies accelerates the diffusion of O ions, thereby promoting the corrosion of grain boundaries.

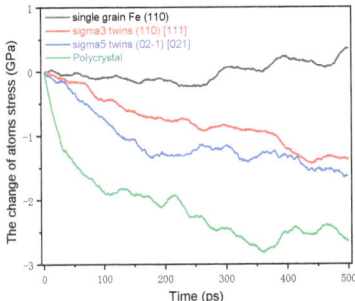

Figure 6. Changes in shear stress of polycrystalline and twin models over simulation time.

The effects of different grain boundaries on the corrosion process were compared by counting the number of dissolved Fe atoms near the grain boundaries and the number of O ions diffused into the Fe substrate. According to Figure 7a, the number of dissolved Fe atoms in polycrystalline and sigma5 twins is the largest due to the early stress release. Within 50 ps, about 250 Fe atoms are dissolved from polycrystalline, and about 100 Fe atoms are dissolved from sigma5 twins. The number of dissolved iron atoms in sigma3 twins and single crystals is roughly the same. Figure 7b shows the number of O ions penetrating into the vicinity of the grain boundaries of the iron substrate. It can be seen that a large number of iron atoms at the grain boundaries of the polycrystalline model are dissolved, and most vacancies are formed on the surface, which causes O ions to penetrate into the substrate quickly. Therefore, the polycrystalline model has the most oxygen atoms penetrating into the substrate. Approximately 100 O ions in sigma5 twins penetrate into the substrate. The number of oxygen atoms in sigma3 twins is 20 more than that in single crystals, but the number of dissolved Fe atoms is the same.

Corrosion at grain boundaries is always accompanied by the massive dissolution of Fe atoms at grain boundaries. Therefore, we simulated the effect of the concentration of vacancies in the Fe substrate on the penetration of O ions in corrosion. Randomly add vacancies in the Fe substrate, and the vacancy concentration is 0%, 5%, and 10%. The result is shown in Figure 8. The concentration of vacancies affects the penetration and depth of O ions penetrate. When the vacancy concentration is 5%, the penetration depth of O ions only increases by 0.25 Å, but when the vacancy concentration rises to 10%, the penetration depth of O ions increases by 1.5 Å. We can conclude that the concentration of vacancies in the Fe substrate reaches a certain level, and O ions can quickly penetrate into the bulk. In intergranular corrosion, the excessive initial stress of the grain boundary causes a large number of Fe atoms at the grain boundary to dissolve to form vacancies, which makes it easier for O ions to penetrate into the substrate to form oxides in accelerated local corrosion.

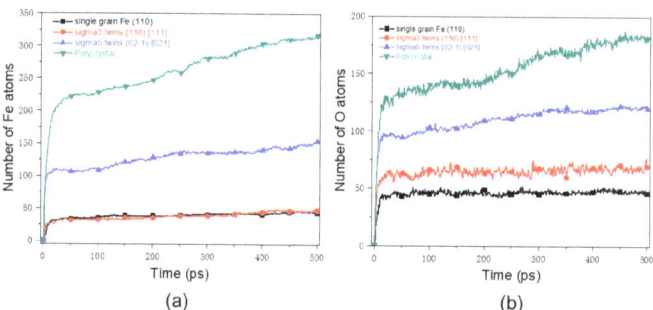

Figure 7. (a) The number of Fe atoms dissolved into H_2O over time. (b) The number of O atoms diffused into the substrate over time.

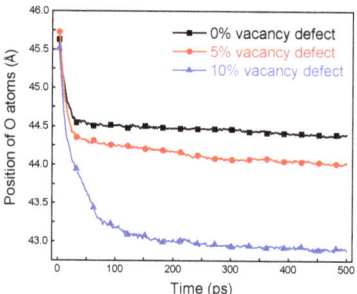

Figure 8. The effect of vacancy concentration on the diffusion of oxygen atoms. The curve represents the average position of oxygen atom diffusion. Simulate by randomly adding vacancies to the Fe substrate proportionally.

4.2. Characterization of Corrosion at the Grain Boundary

The dynamic evolution of the chemical composition of the Fe/water interface system is shown in Figure 9a. In the early 5 ps of corrosion, due to the adsorption and dissociation of many water molecules, a large number of byproducts such as H_3O^+, H_2, OH^-, and H atoms are generated, which means that the early corrosion rate is high. After 10ps, the system enters the slow-corrosion state. After the dissociation of water molecules, O ions diffuse into the Fe substrate, and H ions enter the solution to form H_3O^+. Therefore, as the corrosion progresses, the amount of H_3O^+ gradually increases. Figure 9b shows the changes in the number of water molecules over the simulation time. To avoid interference, we only count the changes of water molecules above the grain boundaries. The changing trend of water molecules in all models is the same. The significant reduction of water molecules in the first 50 ps means a faster corrosion rate. After the initial passivation, the number of water molecules slowly decreases. As the angle of the grain boundary increases, the number of water molecules decreases. It also shows that the existence of grain boundaries will affect the adsorption and dissociation of water molecules and the subsequent penetration of oxygen atoms.

The atomic charge distribution at the end of the pure Fe-H_2O system simulation is depicted in Figure 10. The charge of the Fe atoms at the bottom of the metal substrate has fluctuated around 0 e. The charges of O and H atoms in the solution fluctuate slightly around −0.775 e and 0.320 e, respectively. These atomic charges are consistent with the previous simulation articles [19]. Charge exchange mainly occurs in the electric double layer, where the Fe atom loses electrons, and the charge of the surface Fe atoms rises to nearly 0.7 e. In addition, the charge of oxygen atoms diffused into the Fe substrate decreases to 0.3 e. When hydrogen atoms enter the Fe substrate as interstitial atoms or combine with Fe atoms in the solution, the hydrogen atoms possess a specific negative charge of about

0.2 e. In contrast, the hydrogen atoms in the OH⁻ ions still carry a positive charge. The related research reveals that the H atoms spread into the Fe substrate possess a negative charge [32,33]. Therefore, during the corrosion process, there is a process of redistribution of electric charge on the pure Fe-H_2O system.

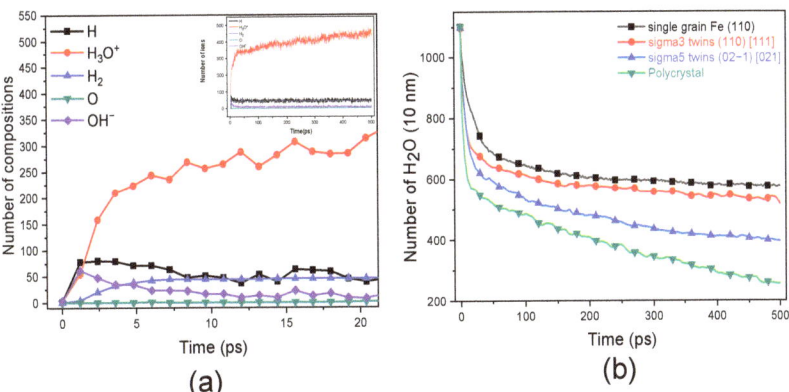

Figure 9. (**a**) The change of ions numbers of polycrystalline during the simulation time and (**b**) the evolution of water molecules in the simulation time.

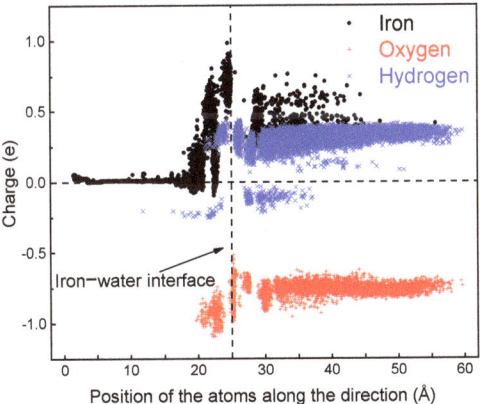

Figure 10. Charge distributions of the polycrystal pure Fe-H_2O system at 500 ps.

The radial distribution functions (RDFs) can study the different Fe oxide phases in simulation can be studied by the radial distribution functions (RDFs). Generally, the partial radial distribution function (RDF), g (r), defined as the probability of an atom at a distance from the origin, which is used to characterize the bonding and the structure of the formed oxide film [34,35]. Figure 11 shows the RDFs for the O-H bond and the Fe-O bond at the end of the simulation. The RDFs were calculated every 0.4 ps and took an average of 450 to 500 ps. It has two prominent peaks for the Fe-O bond length. The first peaks at 1.78 Å correspond to the interaction of Fe ions and OH- ions, but this peak spans from 1.5 Å to 2.5 Å covering the peaks of FeO, Fe_2O_3, and Fe_3O_4. These results show that the oxide film formed on the Fe surface mainly consists of Fe hydroxides and oxides in the corrosion process. The transformation trend is the transformation of Fe hydroxide to Fe oxide.

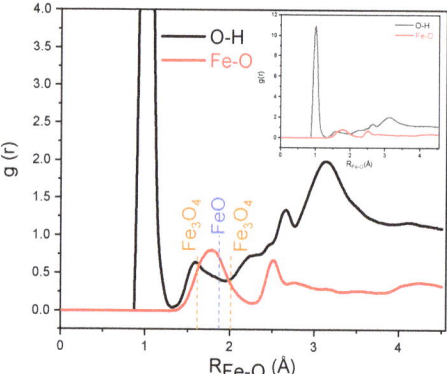

Figure 11. Partial radial distribution functions for the oxide film of polycrystal model at 500 ps.

5. Conclusions

In this paper, the reactive molecular dynamics simulation method studied the effect of different grain boundaries on the corrosion interface process of pure Fe-H$_2$O. Our study shows that the stress release in the high-stress region of the grain boundary during the corrosion process promotes the rapid penetration of O ions into the grain boundary to form oxides, resulting in the occurrence of intergranular corrosion in pure iron. The effect of grain boundaries on intergranular corrosion is revealed from the point of view of simulation, which provides a theoretical explanation for the occurrence of intergranular corrosion in Fe-H$_2$O corrosion in a realistic environment.

Author Contributions: Conceptualization, Z.L. and C.H.; methodology, Z.L. and C.H.; validation, Z.X. and Y.H.; formal analysis, Z.X.; investigation, Z.X.; data curation, Z.X.; writing—original draft preparation, Z.X.; writing—review and editing, Z.X., Y.H., W.H., C.H. and Q.W.; funding acquisition, Q.W. All authors have read and agreed to the published version of the manuscript.

Funding: This work was supported by the research project for Hua-long Pressurized Reactor (HPR1000) in China National Nuclear Corporation.

Data Availability Statement: Data sharing not applicable.

Conflicts of Interest: The authors declare no conflict of interest.

References

1. Sherif, E.-S.M.; Almajid, A.A.; Khalil, K.A.; Junaedi, H.; Latief, F. Electrochemical studies on the corrosion behavior of API X65 pipeline steel in chloride solutions. *Int. J. Electrochem. Sci.* **2013**, *8*, 9360–9370.
2. Čapek, J.; Stehlíková, K.; Michalcová, A.; Msallamová, Š.; Vojtěch, D. Microstructure, mechanical and corrosion properties of biodegradable powder metallurgical Fe-2 wt% X (X= Pd, Ag and C) alloys. *Mater. Chem. Phys.* **2016**, *181*, 501–511. [CrossRef]
3. Kutz, M. *Handbook of Environmental Degradation of Materials*; William Andrew: Norwich, NY, USA, 2018.
4. Galvele, J.; de De Micheli, S. Mechanism of intergranular corrosion of Al-Cu alloys. *Corros. Sci.* **1970**, *10*, 795–807. [CrossRef]
5. Gwathmey, A. Effect of Crystal Orientation on Corrosion. In *The Corrosion Handbook*; Uhlig, H.H., Ed.; John Wiley & Sons, Inc.: Hoboken, NJ, USA, 1948; Volume 33, pp. 157–162.
6. Zou, Y.; Wang, J.; Zheng, Y. Electrochemical techniques for determining corrosion rate of rusted steel in seawater. *Corros. Sci.* **2011**, *53*, 208–216. [CrossRef]
7. Schultze, J.; Davepon, B.; Karman, F.; Rosenkranz, C.; Schreiber, A.; Voigt, O. Corrosion and passivation in nanoscopic and microscopic dimensions: The influence of grains and grain boundaries. *Corros. Eng. Sci. Technol.* **2004**, *39*, 45–52. [CrossRef]
8. You, D.; Pebere, N.; Dabosi, F. An investigation of the corrosion of pure iron by electrochemical techniques and in situ observations. *Corros. Sci.* **1993**, *34*, 5–15. [CrossRef]
9. Jinlong, L.; Hongyun, L. The effects of cold rolling temperature on corrosion resistance of pure iron. *Appl. Surf. Sci.* **2014**, *317*, 125–130. [CrossRef]
10. Bennett, B.W.; Pickering, H.W. Effect of grain boundary structure on sensitization and corrosion of stainless steel. *Metall. Trans. A* **1991**, *18*, 1117–1124. [CrossRef]

11. Lapeire, L.; Lombardia, E.M.; Verbeken, K.; De Graeve, I.; Kestens, L.; Terryn, H. Effect of neighboring grains on the microscopic corrosion behavior of a grain in polycrystalline copper. *Corros. Sci.* **2013**, *67*, 179–183. [CrossRef]
12. Li, D. Electron work function at grain boundary and the corrosion behavior of nanocrystalline metallic materials. *MRS Online Proc. Libr.* **2005**, *887*, 8870503. [CrossRef]
13. Hoffman, E.E.; Lin, A.; Liao, Y.; Marks, L.D. Grain boundary assisted crevice corrosion in CoCrMo alloys. *Corrosion* **2016**, *72*, 1445–1461. [CrossRef]
14. Hu, J.; Wang, C.; He, S.; Zhu, J.; Wei, L.; Zheng, S. A DFT-Based Model on the Adsorption Behavior of H_2O, H^+, Cl^-, and OH^- on Clean and Cr-Doped Fe (110) Planes. *Coatings* **2018**, *8*, 51. [CrossRef]
15. Nunomura, N.; Sunada, S. Density functional theory based modeling of the corrosion on iron surfaces. *Arch. Metall. Mater.* **2013**, *58*, 321–323. [CrossRef]
16. Zheng, Y.-T.; Xuan, F.-Z.; Wang, Z. The Role of Atomic Structures on the Oxygen Corrosion of Polycrystalline Copper Surface. *Procedia Eng.* **2015**, *130*, 1184–1189. [CrossRef]
17. Liu, X.; Kim, S.-Y.; Lee, S.H.; Lee, B. Atomistic investigation on initiation of stress corrosion cracking of polycrystalline Ni60Cr30Fe10 alloys under high-temperature water by reactive molecular dynamics simulation. *Comput. Mater. Sci.* **2021**, *187*, 110087. [CrossRef]
18. Verners, O.; van Duin, A.C.T. Comparative molecular dynamics study of fcc-Ni nanoplate stress corrosion in water. *Surf. Sci.* **2015**, *633*, 94–101. [CrossRef]
19. DorMohammadi, H.; Pang, Q.; Árnadottir, L.; Isgor, O.B. Atomistic simulation of initial stages of iron corrosion in pure water using reactive molecular dynamics. *Comput. Mater. Sci.* **2018**, *145*, 126–133. [CrossRef]
20. DorMohammadi, H.; Pang, Q.; Murkute, P.; Árnadóttir, L.; Burkan Isgor, O. Investigation of iron passivity in highly alkaline media using reactive-force field molecular dynamics. *Corros. Sci.* **2019**, *157*, 31–40. [CrossRef]
21. DorMohammadi, H.; Pang, Q.; Murkute, P.; Árnadóttir, L.; Isgor, O.B. Investigation of chloride-induced depassivation of iron in alkaline media by reactive force field molecular dynamics. *Npj Mater. Degrad.* **2019**, *3*, 19. [CrossRef]
22. Lu, K.; Huo, C.-F.; He, Y.; Yin, J.; Liu, J.; Peng, Q.; Guo, W.-P.; Yang, Y.; Li, Y.-W.; Wen, X.-D. Grain boundary plays a key role in carbon diffusion in carbon irons revealed by a ReaxFF study. *J. Phys. Chem. C* **2018**, *122*, 23191–23199. [CrossRef]
23. Plimpton, S. Fast parallel algorithms for short-range molecular dynamics. *J. Comput. Phys.* **1995**, *117*, 1–19. [CrossRef]
24. Van Duin, A.C.; Dasgupta, S.; Lorant, F.; Goddard, W.A. ReaxFF: A reactive force field for hydrocarbons. *J. Phys. Chem. A* **2001**, *105*, 9396–9409. [CrossRef]
25. Huang, Y.; Hu, C.; Xiao, Z.; Gao, N.; Wang, Q.; Liu, Z.; Hu, W.; Deng, H. Atomic insight into iron corrosion exposed to supercritical water environment with an improved Fe-H_2O reactive force field. *Appl. Surf. Sci.* **2021**, *580*, 152300. [CrossRef]
26. Smith, D.W. A new method of estimating atomic charges by electronegativity equilibration. *J. Chem. Educ.* **1990**, *67*, 559. [CrossRef]
27. Rappe, A.K.; Goddard, W.A., III. Charge equilibration for molecular dynamics simulations. *J. Phys. Chem.* **1991**, *95*, 3358–3363. [CrossRef]
28. McDonald, I. NpT-ensemble Monte Carlo calculations for binary liquid mixtures. *Mol. Phys.* **1972**, *23*, 41–58. [CrossRef]
29. Hoover, W.G. Canonical dynamics: Equilibrium phase-space distributions. *Phys. Rev. A* **1985**, *31*, 1695. [CrossRef]
30. Nosé, S. A molecular dynamics method for simulations in the canonical ensemble. *Mol. Phys.* **1984**, *52*, 255–268. [CrossRef]
31. Assowe, O.; Politano, O.; Vignal, V.; Arnoux, P.; Diawara, B.; Verners, O.; Van Duin, A. Reactive molecular dynamics of the initial oxidation stages of Ni (111) in pure water: Effect of an applied electric field. *J. Phys. Chem. A* **2012**, *116*, 11796–11805. [CrossRef]
32. Das, N.K.; Suzuki, K.; Takeda, Y.; Ogawa, K.; Shoji, T. Quantum chemical molecular dynamics study of stress corrosion cracking behavior for fcc Fe and Fe–Cr surfaces. *Corros. Sci.* **2008**, *50*, 1701–1706. [CrossRef]
33. Wang, H.; Han, E.-H. Ab initio molecular dynamics simulation on interfacial reaction behavior of Fe-Cr-Ni stainless steel in high temperature water. *Comput. Mater. Sci.* **2018**, *149*, 143–152. [CrossRef]
34. Scott, R. *Computer Simulation of Liquids*; JSTOR: New York, NY, USA, 1991.
35. Frenkel, D.; Smit, B. *Understanding Molecular Simulation: From Algorithms to Applications*; Elsevier: Amsterdam, The Netherlands, 2001; Volume 1.

Article

Effects of Point Defects on the Stable Occupation, Diffusion and Nucleation of Xe and Kr in UO$_2$

Li Wang [1], Zhen Wang [2,*], Yaping Xia [3], Yangchun Chen [3], Zhixiao Liu [1], Qingqing Wang [2], Lu Wu [2], Wangyu Hu [1,*] and Huiqiu Deng [3]

1. College of Materials Science and Engineering, Hunan University, Changsha 410082, China; liwang11@hnu.edu.cn (L.W.); zxliu@hnu.edu.cn (Z.L.)
2. The First Sub-Institute, Nuclear Power Institute of China, Chengdu 610041, China; wqq1132675279@163.com (Q.W.); wulu1002@126.com (L.W.)
3. School of Physics and Electronics, Hunan University, Changsha 410082, China; xiayaping@hnu.edu.cn (Y.X.); ychchen@hnu.edu.cn (Y.C.); hqdeng@hnu.edu.cn (H.D.)
* Correspondence: wangzshu@126.com (Z.W.); wyuhu@hnu.edu.cn (W.H.)

Abstract: Xe and Kr gases produced during the use of uranium dioxide (UO$_2$)-fuelled reactors can easily form bubbles, resulting in fuel swelling or performance degradation. Therefore, it is important to understand the influence of point defects on the behaviour of Xe and Kr gases in UO$_2$. In this work, the effects of point defects on the behavioural characteristics of Xe/Kr clusters in UO$_2$ have been systematically studied using molecular dynamics. The results show that Xe and Kr clusters occupy vacancies as nucleation points by squeezing U atoms out of the lattice, and the existence of vacancies makes the clusters more stable. The diffusion of interstitial Xe/Kr atoms and clusters in UO$_2$ is also investigated. It is found that the activation energy is ~2 eV and that the diffusion of the interstitial atoms is very difficult. Xe and Kr bubbles form at high temperatures. The more interstitial Xe/Kr atoms or vacancies in the system, the easier the clusters form.

Keywords: UO$_2$; Xe; Kr; occupation; diffusion; nucleation; molecular dynamics

1. Introduction

With the increasing consumption of energy on Earth, the development of nuclear energy has attracted considerable research attention from all walks of life. Nuclear fission is a critical way to generate clean energy, and uranium dioxide (UO$_2$) is the standard nuclear fuel used in pressurised water reactors [1]. Fission gases, such as Xe and Kr, are among the essential fission products in UO$_2$ fuel, which can exacerbate the fuel swelling, thereby leading to the interaction between the fuel and the cladding [2–6]. As the fission products deposit energy in the surrounding material, point defects that control the microstructural evolution of the fuel occur. The point defects that survive the initial damage from irradiation in nuclear fuel form extended defects, such as vacancy clusters, dislocation loops, and voids [7]. Numerous experimental and modeling studies have been conducted to improve the understanding of the behaviour of Xe and Kr gases [7–14].

Among all volatile fission products, Xe and Kr have the highest concentration and are mainly studied herein. Previous literature mainly focused on stable configurations with a constant Xe-vacancy ratio; for example, Moore et al. [15] found that clusters of Xe atoms are formed by single Xe atoms occupying Schottky positions, which is caused by the supersaturation of Schottky vacancies in UO$_2$. Due to the complexity of the behavioural characteristics of UO$_2$ fuel materials and Xe bubbles, it is difficult to determine the behavioural mechanism of Xe gases. Consequently, several separation effect experiments have been proposed to simplify complex material systems by describing the physical processes of one or more fission gases to elucidate the underlying behavioural mechanisms. Thus, Zhang et al. [16] briefly explained the mechanism of UO$_2$ by simulating molybdenum.

They simulated the stable configuration of Mo by adding Xe atoms and found that Mo was most stable when the Xe-to-vacancy ratio was unity. This study compares the stable occupation of Xe/Kr clusters in perfect and varying defect-containing systems.

The diffusion of the Xe atoms in bulk UO_2 or Xe-vacancy clusters formed by Xe atoms in Schottky vacancies has been studied in previous literature, and even the self-diffusion behaviour of U and O in UO_2 has been studied [17–19]. Yun et al. [20] have investigated the vacancy-assisted diffusion of Xe in UO_2, and calculated the incorporation, binding, and migration energies. They found that the tri-vacancy is a significant diffusion pathway of Xe in UO_2. Lawrence [21] discussed the uncertainty in fission gas diffusion coefficients as a function of temperature. Higher activation energies in computing diffusion are usually compensated by higher pre-exponential factors [22]. The diffusion of cations in UO_2 and other related compounds is very slow, at <10^{15} or <10^{17} cm^2/s [23,24], even at high temperatures from 1800 to 2000 K, which is one of the highest temperatures achieved in crystal correlation experiments. One of the most commonly used models in fuel performance codes was published by Massih and Forsberg [25–27]. Turnbull et al. [7] analysed this model and other models, and then computed the bulk fission gas diffusion rates, which capture both intrinsic and radiation-enhanced diffusion. This model divides the diffusivity into three regimes. Davies et al. [28] experimentally studied the diffusivity of UO_2 at high temperatures (D_1, T > 1650 K) and concluded that its activation energy (E_a) and pre-exponential factor (D_0) were 3.04 eV and 7.6×10^{-10} m^2/s, respectively. This study provides significant guidance for the subsequent diffusion studies of UO_2 by many researchers [29–31]. The in-pile diffusion coefficient of UO_2 is close to the intrinsic diffusion coefficient, so it is considered that the radiation-enhanced diffusion coefficient has high uncertainty [32]. However, due to the complex diffusion of Xe at interstitial sites, the diffusion of interstitial clusters has not presently been well described. Therefore, it is necessary to explore the diffusion behaviour of Xe/Kr clusters at octahedral interstitial sites in UO_2 and to explain the relationship between the interstitial and vacancy diffusion mechanisms.

There are two main nucleation mechanisms of fission gases, such as homogeneous and heterogeneous nucleation [33]. Transmission electron microscopy (TEM) images of UO_2–irradiated bubbles show that they are characterised by their high density and small, almost uniform bubble size. Nelson [34] predicted that the nucleation density of bubbles was almost independent of irradiation temperature and fission rate. Evans [35] observed bubbles in Kr- and Xe-irradiated UO_2 using TEM and found that the threshold temperature for bubble nucleation was in the range of 350 °C–500 °C. Michel et al. [36] found sub-nanometer Xe bubbles in polycrystalline UO_2 under low flux irradiation at 600 °C. Previous studies focused on the vital role of temperature and irradiation dose in the nucleation and growth of bubbles in UO_2, but there are few studies on defect concentration. Hence, studying the formation of Xe/Kr clusters in systems with different defect concentrations is important for subsequent nucleation studies.

Thus, the influence of defects in materials on Xe/Kr gas clusters is worth further investigation. Therefore, it is crucial to investigate the effect of point defects on Xe/Kr clusters in UO_2 to understand the evolution of fuels.

2. Simulation Method

2.1. Interatomic Potential

The interatomic interaction potentials of UO_2 have been widely reported previously. Among them, the potentials reported by Basak et al. [37], Morelon et al. [38], and Cooper et al. [39] are more commonly used. For the further addition of fission gas, Xe, based on UO_2, and UO_2–Xe interatomic interaction potentials have been mainly developed by Geng et al. [40], Chartier et al. [41], Thompson et al. [42], and Cooper et al. [43]. For the UO_2–Kr system, only one interatomic interaction potential developed by Cooper et al. [43] can be presently used. These UO_2–Xe and UO_2–Kr potentials use the Xe and Kr potentials proposed by Tang and Toennies [44].

Herein, the UO$_2$ potential reported by Cooper et al. [39] is adopted to describe the U–U, U–O, and O–O interactions. This potential reproduces a range of thermophysical properties, such as the lattice parameter, bulk modulus, enthalpy, and specific heat at temperatures between 300 and 3000 K, as well as some defect properties in UO$_2$. In addition, this potential's bulk modulus and elastic constant are more accurate and in accordance with experimental values [45]. The Xe–Xe interaction is described by the Tang–Toennies potential [44]. Further, the interactions of Xe–U and Xe–O have been described by Thompson et al. [42], and are very flexible and can be applied to a wide variety of potential forms and materials systems, including metals and EAM potentials. For the UO$_2$–Kr system, the interatomic interaction potential developed by Cooper et al. [43] is adopted.

2.2. MD Simulation Setup

An MD simulation programme, LAMMPS (7Aug19 version) [46], is employed herein for all simulations. Images of atomic configurations were produced with the visualisation tool OVITO (3.5.0 version, Darmstadt, Germany) [46]. The Wigner–Seitz (W–S) cell method determined the type and position of the interstitial atoms or vacancies [47,48]. The time step was set as 0.001 ps for all MD simulations. The temperature was controlled via a Nose/Hoover temperature thermostat, and the periodic boundary condition was used. For static relaxation, the energy minimisation was performed, and the minimisation algorithm was set as the conjugate gradient method (cg). The specific simulation processes for different behaviours differ, and the details of the different simulations are described as follows.

2.2.1. Stable Occupation of Xe/Kr Cluster in UO$_2$-Containing Point Defects

In addition to studying the stable occupation of Xe(Kr) clusters in defect-free bulk UO$_2$, the influence of different defects in UO$_2$ on the stable occupation of Xe(Kr) clusters was studied, such as U, O, UO double, Schottky, and double Schottky vacancies. A cubic box of 25 a_0 × 25 a_0 × 25 a_0 (a_0 is the lattice constant of the UO$_2$ fluorite structure at 0 K) containing 187,500 atoms was used. MD simulation was performed after generating each configuration to equilibrate the system. After energy minimisation, the first atom was introduced into the system. The site with the lowest energy formation was searched through energy minimisation again to determine the stable site of the first atom. Afterwards, the second atom was introduced around the first atom, and the process was repeated to obtain a stable space for the two atoms. Further, the same process was performed until stable positions for six atoms were found successfully.

The formation energy of a Xe/Kr cluster in defect-free UO$_2$ bulk is defined as follows:

$$E^f_{N\ Xe/Kr} = E^{Int}_{N\ Xe/Kr} - E^P - NE^{Xe/Kr} \tag{1}$$

where $E^{Int}_{N\ Xe/Kr}$ is the energy of the UO$_2$ system containing the Xe (or Kr) cluster, E^P is the energy of perfect UO$_2$, N is the number of Xe (or Kr) atoms, and $E^{Xe/Kr}$ is the energy of a single-isolated Xe (or Kr) atom (this value is zero for the interatomic potential under consideration).

The formation energy of a Xe/Kr cluster in defective UO$_2$ is defined as

$$E^f_{N\ Xe/Kr} = E^{Int}_{N\ Xe/Kr} - E^m_{VD} - NE^{Xe/Kr} \tag{2}$$

where $E^{Int}_{N\ Xe/Kr}$ is the total energy of the system with the Xe (or Kr) cluster added on VD (represent different vacancy-type defects), E^m_{VD} is the total energy of systems containing the VD, m is the number of vacancies for VD, N is the number of Xe (or Kr) atoms, and $E^{Xe/Kr}$ is the energy of a single-isolated Xe (or Kr) atom.

The binding energy of an additional X (X = Xe or Kr) atom to a VD-X cluster in UO$_2$ is defined as follows:

$$E^b(X + VD - X\ cluster) = E^f(X) + E^f(VD - X\ cluster) - E^f(X + VD - X\ cluster) \tag{3}$$

where $E^f(X)$ is the formation energy of a Xe (or a Kr) atom located on the most stable interstitial site in bulk UO_2.

2.2.2. Diffusion of Xe/Kr Cluster in UO_2

Generally, diffusion in solids occurs with point defects [18]. The point defect concentration is thermally activated, and it increases as the temperature increases. Migration is also a thermally activated process, accelerated by an increasing temperatures. Hence, the diffusion coefficients and diffusion energy barriers of small Xe and Kr clusters (number of atoms < 6) in UO_2 are calculated by the mean square displacement (MSD) method, which can intuitively reflect the strength of the self-diffusion ability of particles. A box of $10\,a_0 \times 10\,a_0 \times 10\,a_0$ (a_0 is the lattice constant of the UO_2 fluorite structure at different temperatures from 1800 to 2300 K) containing 12,000 atoms was used. The total simulation time is up to 5 ns, with a timestep of 1 fs.

The Arrhenius's equation can express the temperature dependence of the diffusivity,

$$D = D_0 \exp\left(-\frac{E_a}{k_B T}\right) \tag{4}$$

where D is the diffusion coefficient, E_a is the diffusion barrier, T is the temperature, D_0 is a pre-diffusion factor, and k_B is the Boltzmann constant. Taking the logarithms of both sides of the above equation give

$$\ln D = \ln D_0 - \frac{E_a}{k_B T} \tag{5}$$

Therefore, if the $\ln D$ at different simulated temperatures is obtained, E_a can be obtained by linear fitting, whereas the diffusion coefficient D at different temperatures can be obtained by the MSD method:

$$D_T = \frac{(MSD)_T}{2dt} = \frac{<\Delta r(t)^2_T>}{2dt} \tag{6}$$

In the simulation process, a long simulation time and short coordinate position output intervals are used to obtain the atomic coordinate information. The results are obtained by averaging the MSD trajectory segmentation severally.

2.2.3. Nucleation of Xe/Kr Cluster in UO_2-Containing Point Defects

MD simulation was performed to simulate the nucleation process of Xe bubbles. Five systems with sizes of $10\,a_0 \times 10\,a_0 \times 10\,a_0$ (a_0 is the lattice constant of UO_2 fluorite structure at the temperature of 2500 K) were studied, which are mainly perfect bulk without defects, with 1% vacancies concentration, with 2% vacancies concentration, with 1% interstitial concentration, and with 2% interstitial concentration. Then, 2% and 5% concentration of Xe/Kr atoms were added to the different systems, and the Xe/Kr atoms were randomly and uniformly located at octahedral interstitial sites. The specific simulation process is as follows.

After static relaxation of the initial system relaxation of the above configuration with a high temperature, the temperature was raised to 2500 K to accelerate the diffusion of Xe/Kr atoms. Then, the simulation was conducted in the NVT ensemble for 20 ns. Finally, the obtained configuration was subjected to static relaxation to ensure a stable configuration.

3. Results and Discussion

3.1. Stable Occupation of Xe/Kr Cluster in UO_2-Containing Point Defects

By studying the position and energy changes of Xe/Kr atoms in UO_2, it is found that, after adding Xe/Kr atoms into UO_2, the Xe/Kr atoms move to the octahedral interstitial sites after structural optimization. Figure 1a shows that, when a Xe atom is randomly inserted into the bulk UO_2, it moves to the nearest octahedral site. The energy is lowest at the octahedral interstitial site, with a formation energy of 9.79 eV. The phenomenon is the same for the Kr atom, but, because the Kr atom is smaller than the Xe atom, its formation energy is smaller at 8.45 eV. When two interstitial Xe or two interstitial Kr atoms are added into UO_2, after relaxation, the energy of the two atoms forming a dimer is the lowest. The formation energies of Xe_2 and Kr_2 are 16.44 and 15.60 eV, respectively. The stable structure of three interstitial atoms is shown as an equilateral triangle. The formation energies of Xe_3 and Kr_3 are 22.49 and 21.53 eV, respectively. After the addition of the fourth Xe atom and complete relaxation (Figure 1b), the positions of the four atoms appear as triangular cones at different octahedral sites. The first lattice U atom is squeezed out of the cluster, and the Xe atoms cluster around the U vacancy. However, the interstitial U atom moves away from the cluster, and the formation energy of Xe_4 is 27.92 eV. The Kr_4 cluster slightly differs from Xe_4. The four Kr atoms are located on different octahedral gaps, forming a planar quadrilateral. However, no interstitial atoms are squeezed out of clusters, and the formation energy of Kr_4 is 27.38 eV. After inserting the fifth atom, the first lattice U atom in the Kr cluster is squeezed out. After adding the sixth atom, the second lattice U atom is squeezed out of the Xe cluster (Figure 1c). The formation energies of Xe_5, Kr_5, Xe_6, and Kr_6 are 32.19, 32.41, 39.79, and 37.21 eV, respectively.

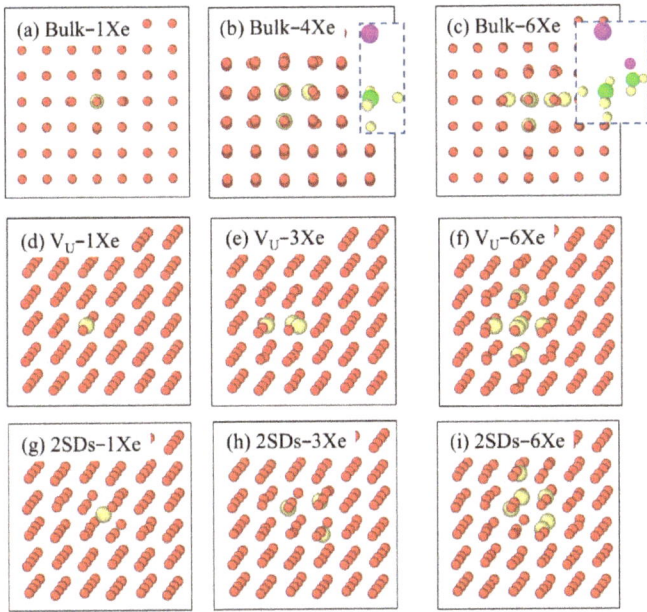

Figure 1. Relaxation configuration diagrams of different UO_2 systems after adding Xe atoms. (**a–c**) are 1, 4, and 6 Xe in the defect-free system, respectively; (**d–f**) are 1, 3, and 6 Xe in the system containing U vacancies, respectively. (**g–i**) are 1, 3, and 6 Xe in the system containing double Schottky vacancies, respectively. The dashed frames are the schematics of the squeezed interstitial atoms. The red, yellow, green, and purple balls represent U atoms, Xe atoms, U vacancies, and U interstitial atoms, respectively. The O atoms are ignored.

Additionally, the stabilities of Xe/Kr atoms in configurations with different defects are compared. As shown in Figure 1d, in the configuration containing a single U vacancy, the addition of one Xe/Kr atom occupies the vacancy. When adding three or six Xe/Kr atoms in this system (Figure 1e,f), the atoms occupy the vacancy and distribute in the octahedral interstitial sites around the vacancy. The configuration of a single O vacancy is consistent with that of a single U vacancy. In the case of the UO double vacancy, one Xe/Kr atom was added to occupy the U vacancy. Two Xe/Kr atoms are evenly distributed into the central region of the two vacancy centres; when multiple atoms are added, they take the central region as the origin and occupy the surrounding octahedral interstitial sites. Figure 1g–i shows that, when there is a double Schottky vacancy, the Xe/Kr atom moves to the position near the central vacancy region. When more than one atom is present in the box, the Xe/Kr atoms are mainly distributed in the central vacancy region or the surrounding octahedral interstitial sites.

Figure 2 shows that, as the number of Xe/Kr atoms increases, the formation energy of the configuration with various defect types increases gradually. The formation energy of the O vacancy was 5 eV larger than that of the U vacancy on average at each stage. The Xe/Kr atom was more accessible to form in the U vacancy than in the O vacancy. Further, Figure 2 shows that the formation energy of Xe/Kr clusters in the six systems can be divided into three layers. The first layer contains the bulk UO_2 and the system with O vacancy. They are characterized by no U vacancy, and the formation energy difference is very small. The second layer contains U, UO double, and Schottky vacancies, which contain only one U vacancy. The third layer contains double Schottky vacancies, which contain two U vacancies. Additionally, the volume of the U vacancy is much larger than that of the O vacancy, which provides more space for clusters. Thus, the stability of Xe/Kr clusters depends on the number of U vacancies. Figure 3 shows that, as the number of Xe/Kr atoms increases, the binding energy of Xe/Kr and cluster decreases and stabilises. The overall trends of adding Xe or Kr atoms are consistent, and the changes caused by different defects in the system are similar. The double Schottky configuration has a more vital ability to adsorb Xe atoms and weaken. When additional atoms are adsorbed to a certain extent, the adsorption capacities of all defective configurations tend to be the same, which mainly depends on the number of vacancy defects contained in the configurations.

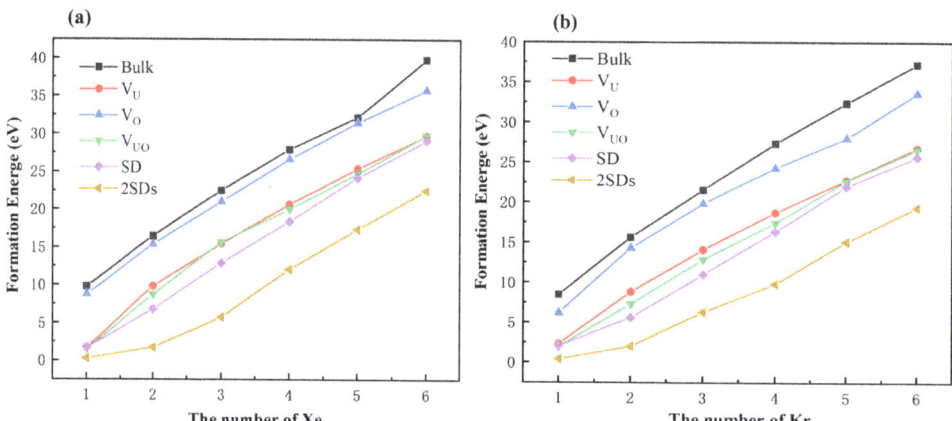

Figure 2. The formation energy (eV) of (**a**) Xe clusters and (**b**) Kr clusters in UO_2 with and without defects as a function of the number of Xe or Kr atoms in the formed cluster.

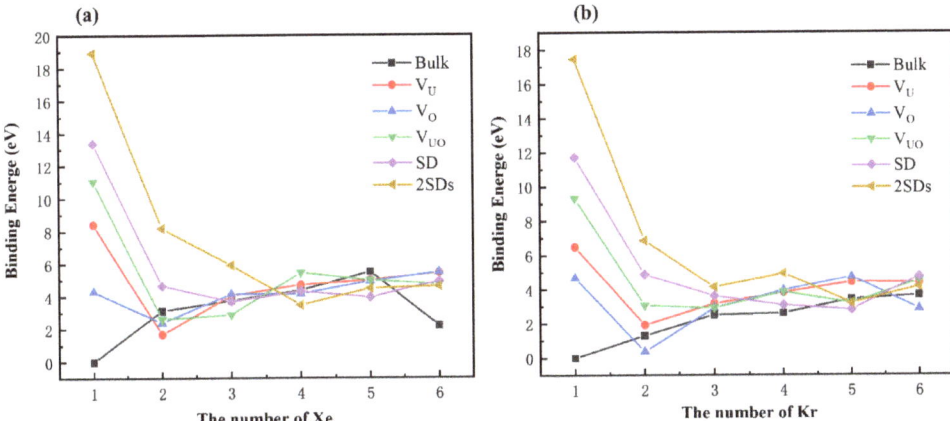

Figure 3. The binding energy (eV) of (**a**) an additional Xe atom and (**b**) an additional Kr atom to a cluster as a function of the number of Xe or Kr atoms in the formed cluster.

3.2. Diffusion of Xe/Kr Cluster in UO$_2$

Based on empirical potential calculations, the diffusivity of Xe/Kr clusters in bulk UO$_2$ has been calculated. Since the first U interstitial atom was excited by adding four Xe atoms or five Kr atoms to the bulk UO$_2$, we studied the Xe/Kr clusters with less than four atoms. Xe usually diffuses due to a vacancy-assisted mechanism. The diffusion of Xe at U, O, UO, one U, two O vacancies, and vacancy clusters (comprising two U vacancies and zero, one, or two O vacancies) has been studied in most studies. Earlier studies concluded that Xe atoms occupied trap sites that contained at least one uranium vacancy and, in many cases, one or two additional oxygen vacancies [49,50]. The conclusion showed that triple vacancy was the main diffusion pathway of Xe in UO$_2$. Previous DFT data [51–53] have shown the activation energies of Xe from 2.87 to 3.95 eV and pre-diffusion factors from 5×10^{-4} m^2/s to 2.9×10^{-12} m^2/s. Due to experimental factors, Lawrence et al. [21] found that the diffusion coefficients between different studies have many orders of magnitude. Herein, the interstitial diffusion mechanism of Xe/Kr was investigated. The simulation estimated the diffusion barrier of Xe atoms as 2.11 eV and the pre-diffusion factor index as 1.8×10^{-5} m^2/s at temperatures between 1800 and 2300 K, and the simulation estimated the diffusion barrier of Kr atoms as 2.31 eV and the pre-diffusion factor index as 0.12×10^{-3} m^2/s. Tables 1 and 2 show the detailed data of the Xe/Kr atom and clusters. Torres et al. [54] calculated the migration energies of Xe/Kr in bulk UO$_2$ by a direct mechanism, and the results were 4.09 and 4.72 eV, respectively.

Table 1. Diffusion energy barrier and diffusion prefactor of small interstitial Xe clusters in UO$_2$.

Number of Xe	Diffusion Energy Barrier (eV)	Diffusion Prefactor (m^2/s)
Xe$_1$	2.11	1.8×10^{-5}
Xe$_2$	2.15	0.35×10^{-5}
Xe$_3$	2.07	0.25×10^{-5}

Table 2. Diffusion energy barrier and diffusion prefactor of small interstitial Kr clusters in UO$_2$.

Number of Kr	Diffusion energy Barrier (eV)	Diffusion Prefactor (m^2/s)
Kr$_1$	2.31	0.12×10^{-3}
Kr$_2$	1.89	0.20×10^{-5}
Kr$_3$	1.95	0.12×10^{-5}

The activation energy of the Xe cluster was ~2 eV, and the diffusion coefficient can be seen in Figure 4, which shows the diffusion coefficient for Xe and Kr clusters in bulk UO_2. Figure 4 illustrates the difficulty of cluster diffusion. It is consistent with the data proposed by Davies et al., indicating that clusters are not easy to diffuse. By analysing the movement of the atoms during migration, we find that, when studying the diffusion of individual atoms, evidently, individual atoms are fast and have a wide range of motion. When studying clusters with two atoms, the atoms move mainly by the rotating bypass method. In the diffusion process, atom A was stapled at random, and then atom B rotated around atom A to find a stable position, and then spread over continuously. When there were three atoms in a cluster, a small cluster was formed with one of the atoms pinned together, and the remaining atoms rotated slightly, causing the whole cluster to move and spread out. These trajectories suggested that the diffusion of interstitial clusters is more complicated and may require more complex conditions. In fact, there are little data on experimental interstitial diffusion.

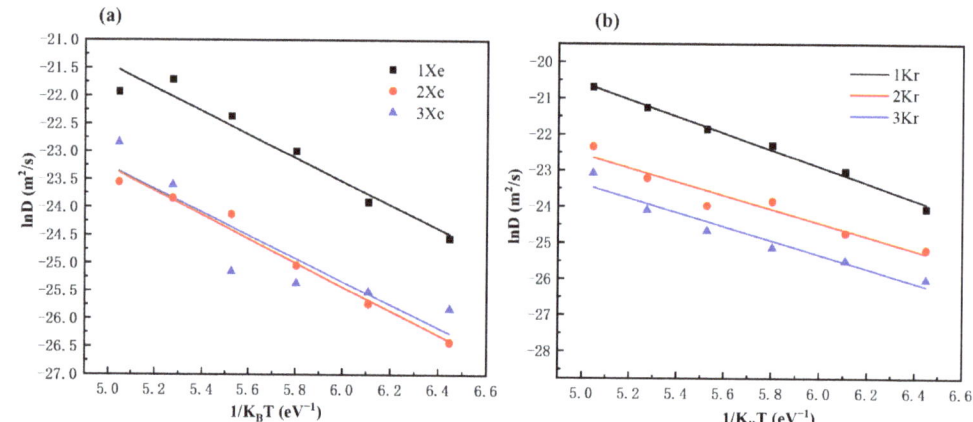

Figure 4. The diffusion coefficient for (**a**) Xe and (**b**) Kr clusters in bulk UO_2. The lines are linear Arrhenius fits.

3.3. Nucleation of Xe/Kr Cluster in UO_2-Containing Point Defects

Here, the Xe/Kr atoms cluster together at high temperatures. A similar phenomenon occurred while studying Mo. Zhang et al. [16] studied the clustering process of Xe atoms dispersed in Mo at high temperatures. They observed the formation of Xe bubbles when the concentration of Xe atoms exceeded the threshold concentration value. In this paper, we studied randomly distributed Xe/Kr atoms at octahedral interspaces in UO_2. Figure 5 shows that, as the relaxation progresses to 5 ns, small Xe clusters form, and then the tiny clusters gradually grow larger by absorbing extra Xe atoms. The clusters are more evident and are larger, almost stable clusters at relaxation to 10 ns. To ensure the stability of clusters, we observed the clustering phenomenon until 20 ns, which was almost not very different from that at 10 ns. The simulation results were consistent with the growth model proposed by Turnbull [55]. The bubbles were heterogeneously nucleated in the wake of fission fragments. They grew by collecting gas by atomic diffusion for a time controlled by a resolution process.

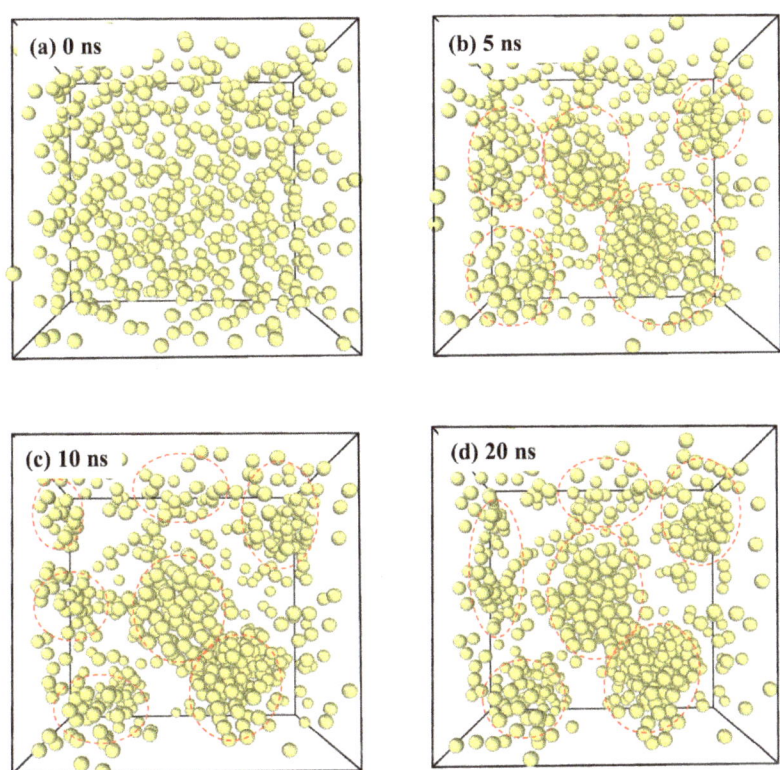

Figure 5. System evolution of Xe clusters at different times at 2500 K (the yellow balls represent Xe atoms, the red circles mark the clusters); (**a**) 0 ns, (**b**) 5 ns, (**c**) 10 ns, (**d**) 20 ns.

In the system with the same defect concentration, the number of clusters formed increases as the interstitial Xe/Kr atomic concentration increases. Figure 6 shows that the number and size of clusters in the system with 5% interstitial Xe atoms added significantly exceeded those with 2% interstitial Xe atoms. The former are more likely to form larger clusters, with a considerable number of clusters over 50 atoms or even over 100 atoms in size. In comparison, the latter are mainly distributed in 2 to 50 atoms.

When there are equal interstitial atom concentrations, the system with more vacancies is more likely to form larger clusters. Similarly, the more interstitial atoms prearranged in the system, the more difficult it is for the Xe/Kr atoms to aggregate during relaxation. The system mainly forms many small clusters ranging in size from 2 to 10 atoms.

The W–S cell method was used to analyse the defect results of the five systems. Three systems were selected for a detailed demonstration, namely the defect-free system (bulk), the system with 1% vacancy concentration (1% vac), and the system with 2% vacancy concentration (2% vac). Figure 7 shows the final distribution of Xe atoms at 5% concentration and the distribution of vacancy atoms and interstitial atoms after defect analysis, respectively. The Xe/Kr atom cluster region overlaps with the position of the vacancy cluster. In other words, the nucleation of the Xe/Kr atom mainly occupied the vacancy position by squeezing out the atoms on the original lattice, which is consistent with the analysis of the stable occupation above. The Xe/Kr atoms squeeze out U atoms to form vacancies, and interstitial atoms moved away from clusters. In addition, Xe/vac is ~1.

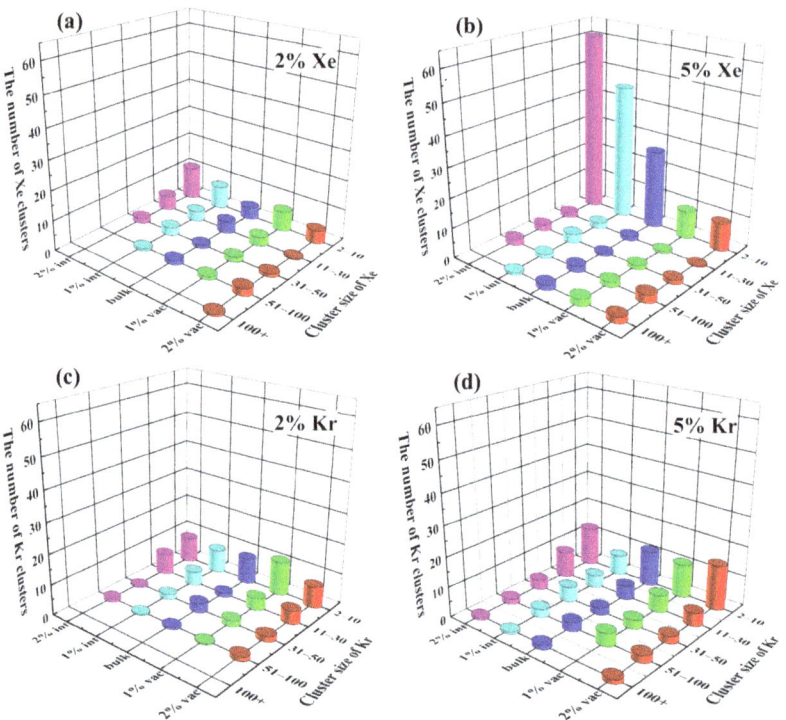

Figure 6. The cluster sizes distribution of (**a**) 2% Xe, (**b**) 5% Xe, (**c**) 2% Kr, and (**d**) 5% Kr in different systems, respectively. The pink, light blue, dark blue, green and red bars are the systems with 2% interstitial concentration, with 1% interstitial concentration, perfect bulk without defects, with 1% vacancies concentration, with 2% vacancies concentration, respectively.

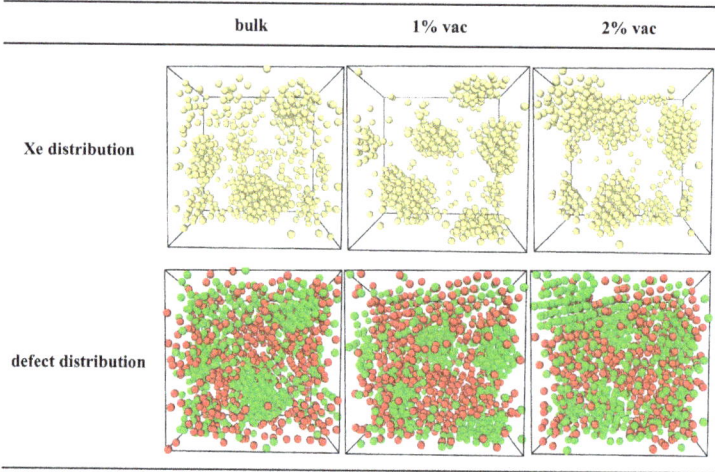

Figure 7. Xe clusters distribution and defects distribution at 5% concentration of Xe atoms in different systems. The green, red, and yellow balls represent the lattice U vacancies, U interstitials, and Xe atoms.

4. Conclusions

In this paper, molecular dynamics simulations were used to study the stable occupancy of Xe/Kr clusters in defect-free configurations and configurations with five different defects. The results show that the system was energetically favourable when Xe/Kr clusters were trapped by vacancies, especially U vacancies, because they can provide a larger space, and the complex of Xe/Kr and vacancies were more stable. The diffusion of Xe/Kr atoms and small clusters at octahedral interstitial sites in bulk UO_2 was also studied. The diffusion of Xe/Kr atoms in bulk UO_2 was relatively complex. The activation energy of Xe/Kr small clusters is ~2 eV, and the Xe/Kr clusters are difficult to diffuse. The nucleation of Xe/Kr in systems containing different defect concentrations was also investigated. At 2500 K, the Xe/Kr atoms dispersed in the system gather into clusters after a 20 ns relaxation. A comparison of the different systems revealed that the vacancy-containing system is more likely to form large clusters than a system containing interstitial atoms.

Author Contributions: Conceptualization, W.H. and H.D.; methodology, L.W. (Li Wang), Z.W., Y.X., Y.C., Z.L., Q.W. and W.H.; validation, Y.X., Z.W., L.W. (Li Wang), Y.C., Z.L., Q.W. and L.W. (Lu Wu); formal analysis, L.W. (Li Wang), Z.W., Y.X., Y.C., Z.L. and Q.W.; investigation, L.W. (Li Wang); resources, Z.W., Z.L. and Q.W.; data curation, Y.C. and W.H.; writing—original draft, L.W. (Li Wang); writing—review & editing, L.W. (Li Wang), L.W. (Lu Wu) and H.D.; supervision, H.D. All authors have read and agreed to the published version of the manuscript.

Funding: This research received no external funding.

Institutional Review Board Statement: Not applicable.

Informed Consent Statement: Not applicable.

Data Availability Statement: Data sharing not applicable.

Conflicts of Interest: The authors declare no conflict of interest.

References

1. Matthews, J.R. Technological problems and the future of research on the basic properties of actinide oxides. *J. Chem. Soc. Faraday Trans. 2* **1987**, *83*, 1273–1285. [CrossRef]
2. Matzke, H. Diffusion processes in nuclear fuels. In *Diffusion Processes in Nuclear Materials*; Agarwala, P., Ed.; Elsevier (North Holland Publishing): Amsterdam, The Netherlands, 1992.
3. Rest, J.; Hofman, G. An alternative explanation for evidence that xenon depletion, pore formation, and grain subdivision begin at different local burnups. *J. Nucl. Mater.* **2000**, *277*, 231–238. [CrossRef]
4. Matzke, H.J. Gas release mechanisms in UO_2—a critical review. *Radiat. Eff.* **1980**, *53*, 219–242. [CrossRef]
5. White, R.; Tucker, M. A new fission-gas release model. *J. Nucl. Mater.* **1983**, *118*, 1–38. [CrossRef]
6. White, R. The development of grain-face porosity in irradiated oxide fuel. *J. Nucl. Mater.* **2004**, *325*, 61–77. [CrossRef]
7. Turnbull, J.; Friskney, C.; Findlay, J.; Johnson, F.; Walter, A. The diffusion coefficients of gaseous and volatile species during the irradiation of uranium dioxide. *J. Nucl. Mater.* **1982**, *107*, 168–184. [CrossRef]
8. He, L.; Pakarinen, J.; Kirk, M.; Gan, J.; Nelson, A.; Bai, X.; El-Azab, A.; Allen, T. Microstructure evolution in Xe-irradiated UO_2 at room temperature. *Nucl. Instrum. Methods Phys. Res. Sect. B Beam Interact. Mater. Atoms* **2014**, *330*, 55–60. [CrossRef]
9. Millett, P.C.; Tonks, M. Meso-scale modeling of the influence of intergranular gas bubbles on effective thermal conductivity. *J. Nucl. Mater.* **2011**, *412*, 281–286. [CrossRef]
10. Chockalingam, K.; Millett, P.C.; Tonks, M. Effects of intergranular gas bubbles on thermal conductivity. *J. Nucl. Mater.* **2012**, *430*, 166–170. [CrossRef]
11. Lucuta, P.; Matzke, H.; Hastings, I. A pragmatic approach to modelling thermal conductivity of irradiated UO_2 fuel: Review and recommendations. *J. Nucl. Mater.* **1996**, *232*, 166–180. [CrossRef]
12. Hoh, A.; Matzke, H. Fission-enhanced self-diffusion of uranium in UO_2 and UC. *J. Nucl. Mater.* **1973**, *48*, 157–164. [CrossRef]
13. Jackson, R.; Catlow, C. Trapping and solution of fission Xe in UO_2.: Part 1. Single gas atoms and solution from underpressurized bubbles. *J. Nucl. Mater.* **1985**, *127*, 161–166. [CrossRef]
14. Djourelov, N.; Marchand, B.; Marinov, H.; Moncoffre, N.; Pipon, Y.; Nédélec, P.; Toulhoat, N.; Sillou, D. Variable energy positron beam study of Xe-implanted uranium oxide. *J. Nucl. Mater.* **2013**, *432*, 287–293. [CrossRef]
15. Moore, E.; Corrales, L.R.; Desai, T.; Devanathan, R. Molecular dynamics simulation of Xe bubble nucleation in nanocrystalline UO_2 nuclear fuel. *J. Nucl. Mater.* **2011**, *419*, 140–144. [CrossRef]
16. Zhang, W.; Yun, D.; Liu, W. Xenon Diffusion Mechanism and Xenon Bubble Nucleation and Growth Behaviors in Molybdenum via Molecular Dynamics Simulations. *Materials* **2019**, *12*, 2354. [CrossRef]

17. Govers, K.; Verwerft, M. Classical molecular dynamics investigation of microstructure evolution and grain boundary diffusion in nano-polycrystalline UO_2. *J. Nucl. Mater.* **2013**, *438*, 134–143. [CrossRef]
18. Vincent-Aublant, E.; Delaye, J.-M.; Van Brutzel, L. Self-diffusion near symmetrical tilt grain boundaries in UO_2 matrix: A molecular dynamics simulation study. *J. Nucl. Mater.* **2009**, *392*, 114–120. [CrossRef]
19. Murphy, S.T.; Jay, E.E.; Grimes, R.W. Pipe diffusion at dislocations in UO_2. *J. Nucl. Mater.* **2014**, *447*, 143–149. [CrossRef]
20. Yun, Y.; Kim, H.; Kim, H.; Park, K. Atomic diffusion mechanism of Xe in UO_2. *J. Nucl. Mater.* **2008**, *378*, 40–44. [CrossRef]
21. Lawrence, G. A review of the diffusion coefficient of fission-product rare gases in uranium dioxide. *J. Nucl. Mater.* **1978**, *71*, 195–218. [CrossRef]
22. Andersson, D.; Garcia, P.; Liu, X.-Y.; Pastore, G.; Tonks, M.; Millett, P.; Dorado, B.; Gaston, D.; Andrs, D.; Williamson, R.; et al. Atomistic modeling of intrinsic and radiation-enhanced fission gas (Xe) diffusion in $UO_{2\pm x}$: Implications for nuclear fuel performance modeling. *J. Nucl. Mater.* **2014**, *451*, 225–242. [CrossRef]
23. Matzke, H. Atomic transport properties in UO_2 and mixed oxides (U, Pu)O2. *J. Chem. Soc. Faraday Trans. 2* **1987**, *83*, 1121–1142. [CrossRef]
24. Sabioni, A.; Ferraz, W.; Millot, F. First study of uranium self-diffusion in UO_2 by SIMS. *J. Nucl. Mater.* **1998**, *257*, 180–184. [CrossRef]
25. Forsberg, K.; Massih, A. Diffusion theory of fission gas migration in irradiated nuclear fuel UO_2. *J. Nucl. Mater.* **1985**, *135*, 140–148. [CrossRef]
26. Forsberg, K.; Massih, A. Fission gas release under time-varying conditions. *J. Nucl. Mater.* **1985**, *127*, 141–145. [CrossRef]
27. Lanning, D.D.; Beyer, C.E.; Painter, C.L. *FRAPCON-3: Modifications to Fuel Rod Material Properties and Performance Models for High-Burnup Application*; Technical Report No. NUREG/CR-6534-Vol. 1; Division of Systems Technology, Office of Nuclear Regulatory Commission, U.S. Nuclear Regulatory Commission: Washington, DC, USA, 1997. [CrossRef]
28. Davies, D.; Long, G. *The Emission of Xenon-133 from Lightly Irradiated Uranium Dioxide Spheroids and Powders*; Technical Report No. AERE-R-4347; Atomic Energy Research Establishment, United Kingdom Atomic Energy Authority: Harwell, England, 1963.
29. Williams, N.R.; Molinari, M.; Parker, S.C.; Storr, M. Atomistic investigation of the structure and transport properties of tilt grain boundaries of UO_2. *J. Nucl. Mater.* **2014**, *458*, 45–55. [CrossRef]
30. Bertolus, M.; Freyss, M.; Dorado, B.; Martin, G.; Hoang, K.; Maillard, S.; Skorek, R.; Garcia, P.; Valot, C.; Chartier, A.; et al. Linking atomic and mesoscopic scales for the modelling of the transport properties of uranium dioxide under irradiation. *J. Nucl. Mater.* **2015**, *462*, 475–495. [CrossRef]
31. Boyarchenkov, A.; Potashnikov, S.; Nekrasov, K.; Kupryazhkin, A. Investigation of cation self-diffusion mechanisms in $UO_{2\pm x}$ using molecular dynamics. *J. Nucl. Mater.* **2013**, *442*, 148–161. [CrossRef]
32. Childs, B. Fission product effects in uranium dioxide. *J. Nucl. Mater.* **1963**, *9*, 217–244. [CrossRef]
33. Olander, D.; Wongsawaeng, D. Re-solution of fission gas—A review: Part I. Intragranular bubbles. *J. Nucl. Mater.* **2006**, *354*, 94–109. [CrossRef]
34. Nelson, R. The stability of gas bubbles in an irradiation environment. *J. Nucl. Mater.* **1969**, *31*, 153–161. [CrossRef]
35. Evans, J.H. Effect of temperature on bubble precipitation in uranium dioxide implanted with krypton and xenon ions. *J. Nucl. Mater.* **1992**, *188*, 222–225. [CrossRef]
36. Michel, A.; Sabathier, C.; Carlot, G.; Kaïtasov, O.; Bouffard, S.; Garcia, P.; Valot, C. An in situ TEM study of the evolution of Xe bubble populations in UO_2. *Nucl. Instrum. Methods Phys. Res. Sect. B Beam Interact. Mater. Atoms* **2012**, *272*, 218–221. [CrossRef]
37. Basak, C.; Sengupta, A.; Kamath, H. Classical molecular dynamics simulation of UO_2 to predict thermophysical properties. *J. Alloy. Compd.* **2003**, *360*, 210–216. [CrossRef]
38. Morelon, N.-D.; Ghaleb, D.; Delaye, J.-M.; Van Brutzel, L. A new empirical potential for simulating the formation of defects and their mobility in uranium dioxide. *Philos. Mag.* **2003**, *83*, 1533–1555. [CrossRef]
39. Cooper, M.W.D.; Rushton, M.; Grimes, R.W. A many-body potential approach to modelling the thermomechanical properties of actinide oxides. *J. Phys. Condens. Matter* **2014**, *26*, 105401. [CrossRef]
40. Geng, H.; Chen, Y.; Kaneta, Y.; Kinoshita, M. Molecular dynamics study on planar clustering of xenon in UO_2. *J. Alloy. Compd.* **2008**, *457*, 465–471. [CrossRef]
41. Chartier, A.; Van Brutzel, L.; Freyss, M. Atomistic study of stability of xenon nanoclusters in uranium oxide. *Phys. Rev. B* **2010**, *81*, 174111. [CrossRef]
42. Thompson, A.E.; Meredig, B.; Stan, M.; Wolverton, C. Interatomic potential for accurate phonons and defects in UO_2. *J. Nucl. Mater.* **2014**, *446*, 155–162. [CrossRef]
43. Cooper, M.W.D.; Kuganathan, N.; Burr, P.A.; Rushton, M.J.D.; Grimes, R.W.; Stanek, C.R.; Andersson, D.A. Development of Xe and Kr empirical potentials for CeO_2, ThO_2, UO_2 and PuO_2, combining DFT with high temperature MD. *J. Phys. Condens. Matter* **2016**, *28*, 405401. [CrossRef]
44. Tang, K.T.; Toennies, J.P. The van der Waals potentials between all the rare gas atoms from He to Rn. *J. Chem. Phys.* **2003**, *118*, 4976–4983. [CrossRef]
45. Padel, A.; De Novion, C. Constantes elastiques des carbures, nitrures et oxydes d'uranium et de plutonium. *J. Nucl. Mater.* **1969**, *33*, 40–51. [CrossRef]
46. Stukowski, A. Visualization and analysis of atomistic simulation data with OVITO—The Open Visualization Tool. *Model. Simul. Mater. Sci. Eng.* **2009**, *18*, 015012. [CrossRef]

47. Kittel, C. *Introduction to Solid State Physics*, 8th ed.; John Wiley & Sons, Inc.: Hoboken, NJ, USA, 2005.
48. Nordlund, K.; Averback, R.S. Point defect movement and annealing in collision cascades. *Phys. Rev. B* **1997**, *56*, 2421–2431. [CrossRef]
49. Ball, R.G.J.; Grimes, R.W. Diffusion of Xe in UO_2. *J. Chem. Soc. Faraday Trans.* **1990**, *86*, 1257–1261. [CrossRef]
50. Catlow, C.R.A. Fission gas diffusion in uranium dioxide. *Proc. R. Soc. Lond. A. Math. Phys. Eng. Sci.* **1978**, *364*, 473–497. [CrossRef]
51. Miekeley, W.; Felix, F. Effect of stoichiometry on diffusion of xenon in UO_2. *J. Nucl. Mater.* **1972**, *42*, 297–306. [CrossRef]
52. Cornell, R.M. The growth of fission gas bubbles in irradiated uranium dioxide. *Philos. Mag.* **1969**, *19*, 539–554. [CrossRef]
53. Kaimal, K.; Naik, M.; Paul, A. Temperature dependence of diffusivity of xenon in high dose irradiated UO_2. *J. Nucl. Mater.* **1989**, *168*, 188–190. [CrossRef]
54. Torres, E.; Kaloni, T. Thermal conductivity and diffusion mechanisms of noble gases in uranium dioxide: A DFT+U study. *J. Nucl. Mater.* **2019**, *521*, 137–145. [CrossRef]
55. Turnbull, J. The distribution of intragranular fission gas bubbles in UO_2 during irradiation. *J. Nucl. Mater.* **1971**, *38*, 203–212. [CrossRef]

Article

Molecular Dynamics Simulations of Xe Behaviors at the Grain Boundary in UO$_2$

Yaping Xia [1], Zhen Wang [2,*], Li Wang [3], Yangchun Chen [1], Zhixiao Liu [3], Qingqing Wang [2], Lu Wu [2] and Huiqiu Deng [1,*]

[1] School of Physics and Electronics, Hunan University, Changsha 410082, China; xiayaping@hnu.edu.cn (Y.X.); ychchen@hnu.edu.cn (Y.C.)
[2] The First Sub-Institute, Nuclear Power Institute of China, Chengdu 610041, China; wqq1132675279@163.com (Q.W.); wulu1002@126.com (L.W.)
[3] College of Materials Science and Engineering, Hunan University, Changsha 410082, China; liwang11@hnu.edu.cn (L.W.); zxliu@hnu.edu.cn (Z.L.)
* Correspondence: wangzshu@126.com (Z.W.); hqdeng@hnu.edu.cn (H.D.)

Abstract: In this study, we investigated the behavior of xenon (Xe) bubbles in uranium dioxide (UO$_2$) grain boundaries using molecular dynamics simulations and compared it to that in the UO$_2$ bulk. The results show that the formation energy of Xe clusters at the Σ5 grain boundaries (GBs) is much lower than in the bulk. The diffusion activation energy of a single interstitial Xe atom at the GBs was approximately 1 eV lower than that in the bulk. Furthermore, the nucleation and growth of Xe bubbles in the Σ5 GBs at 1000 and 2000 K were simulated. The volume and pressure of bubbles with different numbers of Xe atoms were simulated. The bubble pressure dropped with increasing temperature at low Xe concentrations, whereas the volume increased. The radial distribution function was computed to explore the configuration evolution of Xe bubbles. The bubble structures in the GB and bulk material at the same temperature were also compared. Xe atoms were more regular in the bulk, whereas multiple Xe atoms formed a planar structure at the GBs.

Keywords: UO$_2$; grain boundary; Xe bubble; molecular dynamics

1. Introduction

Uranium dioxide (UO$_2$) has been widely used as fuel for nuclear reactors owing to its properties, such as there being no specific deformation when it is strongly irradiated, having an unchanged lattice structure at high temperatures, being non-volatile and being chemically unreactive with water [1]. During the operation of nuclear reactors, the nuclear fuel elements are subjected to a harsh working environment and numerous radioactive fission products are produced during the fission of the fuel assembly in the reactor core. With the development of burnup, solid and gas fission products are produced in the fuel elements and their volume is greater than that of the material before fission. The volume of the fuel element increases with the development of burnup, which is called irradiation swelling. The radiation swelling of nuclear fuel induces interactions between fuel pellets and cladding, resulting in radial deformation and transverse tension of the cladding tube, causing damage to the cladding tube, which seriously threatens the safe operation of the reactor [2]. Swelling caused by solid fission products is simple and increases linearly with burnup; the behavior of gas fission products is complex and this field has not been extensively studied.

Due to the numerous radiation and structural changes experienced during the life of UO$_2$ and storing, understanding and controlling the microstructural changes requires a comprehensive approach that considers all aspects of the material's behavior, from basic radiation damage processes to longer-term changes in the material microstructure. To better understand the fission gas behavior, such as microstructural changes and swelling, many experimental and theoretical studies have been conducted [3–8]. It is difficult to analyze the behavior of xenon (Xe) atoms in UO$_2$ through experiments [9–12] because the growth mechanism of Xe bubbles and the entire physical process are not well understood. Therefore, intragranular rare bubbles under dynamic conditions, cannot be studied through experiments, making computational simulations the only choice. Niemiec et al. proposed a basic evolution equation describing the kinetics of the nucleation and growth phase transitions to study phase transformation or microstructure formation kinetics in physical systems originally composed of several grains [13]. Atomistic simulations are vital to a provide good insight into the atomic structure and damage mechanisms.

To date, all simulations have been conducted in monocrystals. Gadomski et al. [14] studied the kinetic anomalies occurring in nucleation and growth phenomena in complex systems, such as polycrystalline partly ordered alloys, quasicrystal line assemblies, and mesomorphs, to understand the kinetics of the evolution of the microstructure and the system during growth. UO$_2$ pellets are manufactured through traditional powder metallurgical processes; hence, they are polycrystalline materials composed of particles with a diameter of approximately 10 µm. The opening at the grain boundaries (GBs) is a crucial phenomenon since the energy of defects decreases near the GB and fission gas, which causes aggregation at the GB, and can be released to the outside of the fuel. As a result, the properties of UO$_2$ GBs at the atomic scale have been studied [15–19]. In this study, we compare the early behavior of Xe bubbles in the GBs and the entire block of UO$_2$. Previous numerical studies on this subject have mainly focused on metal systems, such as metallic tungsten [20–22]. Herein, we focus on the behavior of Xe atoms at the GBs and the bulk UO$_2$, including the migration of a single Xe interstitial atom, the nucleation and growth of Xe bubbles and the bubble pressure and expansion associated with bubble growth. This study serves as a good reference for higher length scale simulation models.

2. Simulation Method
2.1. Interatomic Potential

Molecular dynamics (MD) simulation is an effective means of studying the microworld, providing insight into the atomic structure and the mechanism of bubble growth. An MD simulation program, LAMMPS [23], was employed for all simulations in this study. However, the accuracy of MD simulations depends largely on the potential function of the atoms used to support the simulation.

Yang et al. studied the influence of different potential functions on the growth of high-pressure Xe bubbles [3]. With the continuous addition of Xe atoms, the pressure of the bubbles is divided into two stages. In stage I, where the bubble pressure is monotonically increased, the bubble characteristics and the microstructure evolution of UO$_2$ are relatively independent with the interatomic potential used; thus, the results of the five potential functions are approximately the same. In stage II, the bubble pressure is released and then fluctuates and the volume and pressure of the bubbles, as well as the evolution of the bubble configuration and the UO$_2$ matrix as a function of the number of Xe atoms, highly depend on the potentials used [3]. In addition, the formation energy of Xe at the same location in UO$_2$ varies significantly with the potential of UO$_2$ and Xe–UO$_2$ [3,24,25].

To ensure the accuracy of MD simulations herein, we evaluated different sets of potential functions and compared them with density functional theory (DFT) or DFT + U data to select a better comprehensive potential function for adding Xe to UO$_2$ [26–30].

The first and second sets of potentials have the same UO$_2$ potential and were developed by Basak et al. [31]. However, IPR [27] and Geng [32] potentials were used to interact with Xe–UO$_2$. In the third and fourth sets of potentials, the potential reported by Morelon [33] was used for the UO$_2$ matrix, whereas those reported by Chartier [29] and IPR [27] were used to describe the Xe–UO$_2$ interaction. In the fifth and sixth sets of potentials, the CRG potential [25] was used as a potential function of UO$_2$ and the interaction of Xe–UO$_2$ was described by Cooper [34] and IPR [27]. Therefore, the six sets of potentials are called Basak/IPR, Basak/Geng, Morelon/IPR, Morelon/Chartier, CRG/Cooper and

CRG/IPR, respectively. Among them, three potential functions of UO$_2$, are paired with different potential functions of Xe–UO$_2$, because the interactions between Xe–U and Xe–O are determined in Xe–UO$_2$. Thus, even in the case of the same UO$_2$ potential function, we simulated it with different Xe–UO$_2$ potential functions and the results are different.

We can determine the accuracy of using interatomic potentials by comparing them with the energetics of relatively small defects calculated using MD [3]. We developed several physically relevant benchmarks to compare the six sets of potentials on an equal basis. We placed Xe at different sites in UO$_2$ (small, intermediate and large vacant sites). For small incorporation, we used the following defects: Xe in an interstitial (Int$_{Xe}$) and Xe on a uranium site with uranium in the neighboring interstitial (Sub$_{Xe}$-Int$_U$). The formation energies of the intermediate-sized incorporation sites are a Xe atom located at a U vacancy (Xe in Vu), an O vacancy (Xe in Vo), two nearest U and O vacancies (Xe in Vuo), or the three configurations of Schottky defect clusters (Xe in SD). Finally, we calculated the defect energetics of a single large incorporation site with Xe in a double Schottky defect cluster (Xe in 2SDs). To directly compare the MD values, a $2 \times 2 \times 2$ supercell of the cubic fluorite unit cell (96 atoms for stoichiometric UO$_2$) with a lattice constant of 5.454 Å was used in the molecular statics calculations. The simulation results are compared with the data in Table 1 [3]. The difference in energy is small, the maximum error value is 0.28 eV and the difference is 0.1 eV. The energy difference may be attributed to the difference in the machines used. However, in general, the results show that the potentials used are adequate. Then, through the MD simulations, we used a larger supercell (12,000 atoms) for better convergence to calculate the energies of Xe atoms at different sites in the UO$_2$ lattice. In addition to the formation energies, we calculated the migration energy of interstitial Xe ($E^m_{Int_{Xe}}$). The results are compared with selected DFT and DFT + U data in Table 1. The results obtained using the CRG/IPR potentials show the best consistency with the DFT results; thus, we chose CRG/IPR as the potential function for adding Xe to UO$_2$ for our subsequent simulations.

Table 1. Comparison of energy of xenon (Xe) point defects in uranium dioxide (UO$_2$) bulk calculated using various interatomic potentials at 0 K. Charge corrections for charge defects were not considered in the calculations herein.

	Basak/Geng	Morelon/Chartier	Basak/IPR	Morelon/IPR	CRG/IPR	CRG/Cooper	DFT
$E^f_{Int_{Xe}}$	22.73	11.94	9.68	8.32	9.79	11.70	9.73 [27] 9.71 [28] 9.7–12.0 [26]
$E^f_{Xe\ in\ Vu}$	5.76	5.31	1.84	0.70	1.72	3.60	2.0–5.8 [26] 1.95 [28]
$E^f_{Xe\ in\ Vo}$	20.55	9.31	8.41	7.17	7.96	8.26	7.5–9.1 [26] 7.85 [28]
$E^f_{Xe\ in\ V_{uo}}$	5.32	4.98	1.52	0.61	1.56	2.96	1.6 [26] 1.55 [28]
$E^f_{Sub_{Xe}-Int_U}$	20.77	Unstable	11.89	10.76	10.29	12.27	11.33 [29]
$E^f_{Xe\ in\ SD}$ (1)	4.82	4.33	0.91	0.48	0.95	1.99	1.06 [27] 1.2 [29]
$E^f_{Xe\ in\ SD}$ (2)	4.92	4.77	1.53	0.67	1.68	3.02	1.83 [27] 1.8 [29]
$E^f_{Xe\ in\ SD}$ (3)	5.02	4.70	1.81	0.65	1.89	3.45	1.94 [27] 2.3 [29]
$E^f_{Xe\ in\ 2SDs}$	3.46	2.79	0.19	0.16	0.18	0.59	0.27 [27]
$E^m_{Int_{Xe}}$	0.76	0.84	5.54	3.28	4.32	1.91	4.48 [30]

2.2. Model Building

The structure of GBs is essential for predicting the behavior of Xe bubbles. Up to now, most theoretical studies on the properties of GB have been based on molecular statics (MS) or DFT [35,36] without considering the influence of temperature. However, experimental observations have shown that GB microstructures and corresponding properties reflect the effect of temperature on GBs to a certain extent [37–39] because temperature changes may cause the relative sliding of grains at GBs [17,19]. Therefore, heat treatment of GBs is a necessary process to expand the understanding of GB. Hong et al. [40] reported that high temperature can cause complex ion transitions in some GBs. In this study, the symmetrically tilted GB, Σ5(310)/[100], in UO$_2$ was used as a GB model because the Σ5 symmetric tilt GB has been extensively studied both experimentally and theoretically [17,41–43]; thus, we could compare our results with the available data.

In this study, the symmetrically tilted GB Σ5(310)/[100] in UO$_2$ were formed by rotating perfect UO$_2$ crystals. Symmetrically tilted GB means that the crystals on both sides of the GB are tilted towards each other and the interface of the GB is symmetrical to the two grains. The specific process of constructing the GB is as follows. First, half of the perfect UO$_2$ crystal simulation box is rotated along the axis by half of the orientation difference angle. Then, the second crystal is constructed through mirror inversion, taking the first crystal as a reference. However, this method causes an overlap of the atoms at the crystal interface. Thus, we removed the overlapped interface atoms to maintain a neutral charge.

The annealed structures of the Σ5(310) GB based on MD are shown in Figure 1 and a common neighbor analysis was performed to identify the GB region and bulk region. The atomic configuration was viewed using OVITO software [44]. At 1000 K, the structure of the interface is conserved with the repetition of a triangular-like pattern (depicted by lines in Figure 1); however, at 2000 K, a structural change occurs with the boundary showing reduced diamond shapes arranged end-on-end. Previous simulation results show that, with an increase in temperature, the disorder degree of GB increases [45]; e.g., at 2000 K, the Σ5(310) GB has a more distorted triangular pattern [19].

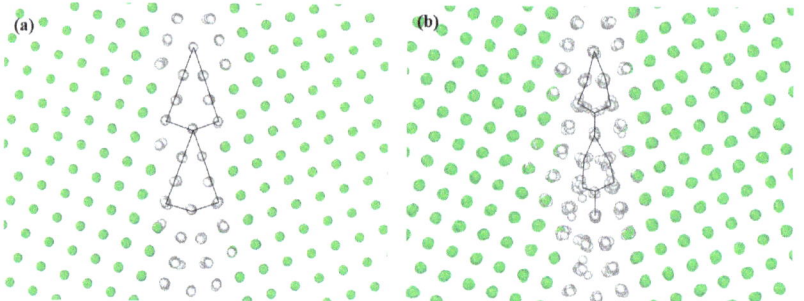

Figure 1. Structures of grain boundaries (GBs) simulated at (**a**) 1000 K and (**b**) 2000 K. Oxygen atoms are removed for clarity. The atoms in the GB and bulk regions are colored white and green, respectively.

2.3. MD Simulation Setup

For all MD simulations, the time step was set as 0.001 ps. For static relaxation, energy minimization was performed and the minimization algorithm was set as the conjugate gradient method (cg), with the stopping criteria for energy tolerance of 10^{-25} s^{-1} and a force tolerance of 10^{-25} eV/Å.

We studied the characteristics of Xe atoms in UO$_2$ GB, including the formation energy of a single Xe atom in the GB, the migration of a single Xe atom in UO$_2$ GB and the nucleation of Xe bubbles, and compared them with those in the bulk. The specific simulation processes for various characteristics are different. Thus we constructed two simulation boxes of different sizes: one contains 38,290 atoms, with a size of 10 a_0 × 20 a_0 × 10 a_0 (a_0 is the lattice constant of UO$_2$) and was used to calculate the migration of a single Xe interstitial atom at the GB, and periodic boundary conditions were used in three directions of the box; the second is a 25 a_0 × 40 a_0 × 25 a_0 large system simulation box and was used to simulate the nucleation and growth of Xe bubbles and the energy of Xe atoms at the GB, and periodic boundary conditions were used in three directions of the box. Similarly, in the bulk, a cubic box of 10 a_0 × 10 a_0 × 10 a_0 containing 12,000 atoms was used with periodic boundary conditions to calculate the migration of a single Xe interstitial atom. The other system is 25 a_0 × 25 a_0 × 25 a_0 with 187,500 atoms and was used to simulate the nucleation and growth of Xe bubbles and the energy of Xe atoms. We performed the same simulation process at the GB and in the bulk to better compare the differences in the behavior of Xe bubbles. A cutoff distance of 1.1 nm was adopted and used for all potentials used in this study.

To determine the stable configuration of Xe atoms in the GB, the stable structures of Xe clusters with different sizes (the number of atoms less than six) were obtained by adding Xe atoms gradually and performing MS simulations for each configuration.

After determining the stable Xe site in the GB, the diffusion mechanism of a single Xe interstitial atom at the GB was studied. The migration of a single Xe atom at the GB was studied using MD and the nudged elastic band (NEB) methods [46–48]. Then, the mean square displacement (MSD) method was used to calculate the diffusion coefficient and diffusion energy barrier of a single Xe atom at the GB.

Finally, to study the evolution mechanism of Xe bubbles, the microcanonical NVT (constant volume and temperature) ensemble was used to simulate Xe bubbles in UO_2 GB. To describe the bubble characteristics during bubble evolution, the volume V and pressure P of the bubble after inserting each Xe atom were calculated by MD. The number of Xe atoms in the bubble is too small (only 50 Xe atoms) to obtain a good statistical result. V is the sum of the volumes of all Xe atoms, which were calculated using the Voronoi technology implemented in LAMMPS [49] and the pressure of the Xe bubble was calculated from the sum of the diagonal component of the atomic stress tensor and the bubble volumes as follows [8,50]:

$$P = -\frac{1}{3V}\sum_i^n [S_{11}(i) + S_{22}(i) + S_{33}(i)] \qquad (1)$$

where n is the number of Xe atoms in the bubble, $S_{\alpha\alpha}(i)$ is the diagonal component of the stress tensor for atom i, V is the volume of all Xe atoms.

Before the first Xe atom was randomly introduced into the simulation box, energy minimization was performed and the minimization algorithm was set using the conjugate method (cg), with the stopping criteria for energy tolerance and force tolerance of 10^{-25}. Then, the first Xe atom was randomly added to the system. The initial system with Xe atoms was minimized to avoid long-distance movements of inadvertently overlapping atoms and the temperatures were set to 1000 K and 2000 K, respectively. The velocities of atoms were set and an additional 10 ps simulation was performed using microcanonical NVT (constant volume and temperature). Following this thermal equilibration (i.e., no change in the various properties of the system over time), Xe atoms were sequentially inserted one by one every 10 ps into the Xe bubble center of mass until 50 Xe atoms were contained in the bubble. After inserting each Xe atom, to avoid introducing artificial energy into the system resulting from atoms placed too close together, the system energy was minimized using the conjugate gradient (cg) and then simulated under NVT conditions for 10 ps at 1000 K and 2000 K.

3. Results

3.1. Formation Energy and Diffusion Behavior of a Single Xe Atom at the GB

The interaction between Xe atoms and the microstructure of UO_2 fuel is key in fission-gas release. To simulate the redistribution of fission gas atoms in the UO_2 microstructure, the interaction range between fission gas atoms and GB atoms must be determined first, to determine the stable structure of migrated Xe atoms at the GB. We calculated the formation energy of a single Xe atom at different positions from the Σ5(310) GB.

There are two distinct regions of a Xe interstitial atom near the GB, as indicated by E in Figure 2. The first region is located at a distance of more than 1.5 a_0 for interstitial Xe, in which the energy is almost equal to the calculated formation energy of Xe interstitial atoms in the bulk (9.89 eV), indicating that the driving force for the segregation of Xe atoms to GB can be ignored. In the second region, where the distance is within 1.5 a_0 from the GB, E decreases, having a maximum value of approximately 6.88 eV, which indicates that interstitial Xe can be absorbed by GB. Thus, the interaction range between Xe interstitial atom and GB is approximately 1.5 a_0 from the plane of GB. Figure 2 also shows that the lowest energy point for interstitial Xe is located at the GB. Thus, it is energetically favourable for interstitial Xe to be absorbed by the GB region.

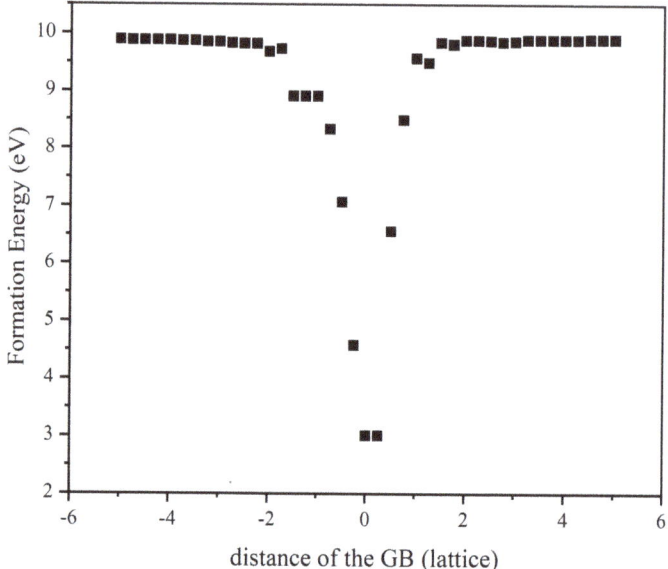

Figure 2. Formation energy of interstitial Xe as a function of the distance from the Σ5(310) GB.

The behavior of fission gas in nuclear fuel is the main factor that determines the change in radiation swelling with fuel consumption. Therefore, it is necessary to study the migration of fission gas in UO_2 GB.

Herein, we calculated the migration of an interstitial Xe atom in UO_2 GB using NEB, i.e., a direct hopping mechanism from one octahedral site to another. NEB showed an energy barrier of 3.87 eV. Thus, whether migration is along or perpendicular to the GB direction, the calculated energy barrier is high. These results suggest that the migration of Xe between the two interstitial sites is not the main mechanism due to the higher energy barrier [51]. The relatively high energy barrier is mainly attributed to the lattice deformation caused by Xe movement between two adjacent interstitial sites, indicating that the octahedral interstitial position in UO_2 is highly strained and, therefore, energetically unfavourable [52,53].

In addition to NEB, we employed the MSD method to calculate the diffusion coefficients and diffusion energy barriers of a single Xe atom at the GB. A long simulation time and a short distance between the position were employed to obtain information on the atomic position. The results were obtained by dividing the tracks of the MSD and averaging them several times. As shown in Figure 3, we calculated the diffusion coefficient and diffusion energy barrier separately at the GB and in the bulk. The diffusion energy barrier of the Xe atoms at the GB was 1.40 eV and that in the bulk was 2.46 eV. The magnitude of the preexponential factor was used to describe the ease of diffusion of a Xe atom. We conclude that Xe atom diffuse more easily at the GB than in the bulk.

The diffusion barrier calculated using MSD and the migration barrier calculated using NEB are different. When NEB was used, the single Xe atom showed a high energy barrier in the interstitial site, which is consistent with that of the bulk. However, the calculation results with MSD are different. This is because Xe may not diffuse directionally in UO_2. The temperature accelerated dynamics simulations revealed that the dynamics of defect clusters strongly depends on their size and the diffusion direction is not one-dimensional [54].

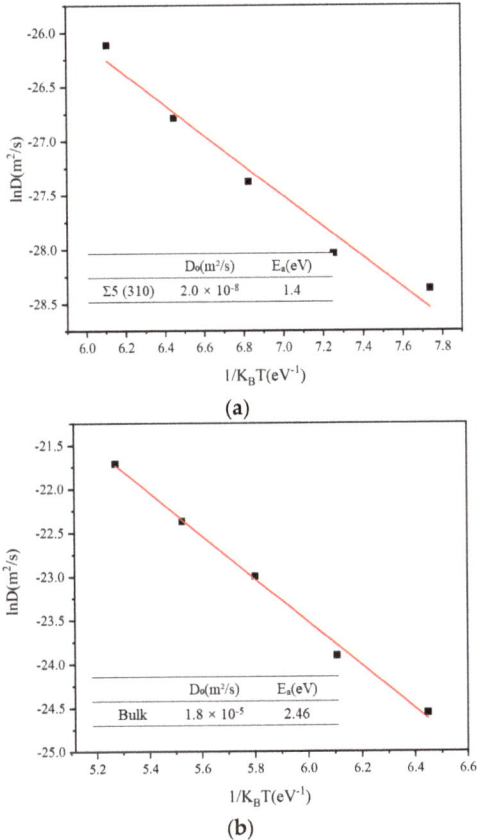

Figure 3. Diffusion coefficients of Xe (**a**) at the Σ5 GB and (**b**) in the bulk UO$_2$. The lines are linear Arrhenius fits.

3.2. Formation Energies of Small Xe Clusters at the GB

We calculated the formation energies of Xe atoms and atomic clusters at the GB and compared them with those in the bulk. The formation energies of Xe atoms at the GB are shown in Table 2. The formation energy of the Xe bubble is defined as:

$$E^f(n) = E_{bubble}(n) - E_{bulk} - nE_{Xe}^{atom} \tag{2}$$

where $E_{bubble}(n)$ is the total energy of the simulation supercell containing n Xe atoms in the bubble, E_{bulk} is the total energy of the supercell without bubbles and E_{Xe}^{atom} is the isolated Xe atom energy, which was set to zero. As shown in Table 2, the energy of a single Xe interstitial atom at the GB was 1.3 eV, whereas that in the bulk it was 9.79 eV. The total energy difference (ΔE) was 8.49 eV. As the number of Xe atoms increases, ΔE increases. Whether at the GB or the bulk, the formation energy increases with an increase in the number of Xe atoms. At all stages, the formation energy of Xe clusters at the GB is smaller than that in the bulk, indicating that Xe bubbles are easier to form at the GB than in the bulk. This is because Xe atoms are at the GB plane, which improves the energy compared to that in the bulk. Similar to previous results [8], at smaller bubble sizes (1–5 Xe atoms), the Xe bubble nucleus at the GB has a much lower formation energy than that of bubbles in the bulk with a similar size.

Table 2. Formation energies of Xe atoms at the GB and in the bulk UO_2.

N	1	2	3	4	5	6
$E^f_{\Sigma5}$ (eV)	1.3	3.8	6.34	10.91	13.41	15.59
E^f_{Bulk} (eV)	9.79	16.44	22.49	27.92	32.19	39.79

3.3. Nucleation of Xe Bubbles at the GB

Early-stage nucleation and growth of Xe bubbles were simulated using the MD method. We simulated the nucleation and growth of Xe bubbles at 1000 and 2000 K. When the system reached a thermodynamic equilibrium after relaxing the system at a given temperature and volume, Xe atoms were regularly added to the bubble. For each Xe atom added, the system needed 10 ps to relax, to ensure there was no change in the properties of the system over time, i.e., the equilibrium state was reached. Thus, the Xe bubble growth was simulated until the bubbles could contain 50 Xe atoms.

Figure 4 lists the differences in the nucleation configurations of Xe bubbles at the Σ5(310) GB and the bulk. At 2000 K, Xe atoms were more regular in the bulk (Figure 4b). This is consistent with a previous report [3], which showed that Xe bubbles evolve into a glassy/amorphous state, a nearly face-centred cubic (FCC) solid structure and subsequently to a high density amorphous or glassy state. At the GB, Xe atoms are absorbed by the GB and multiple Xe atoms form a planar structure, which is similar to the case of rhenium (Re) atoms in metals [55]. The results are consistent with the figure showing the formation energies of an interstitial Xe at the GB. However, comparing the configurations at 1000 K, as shown in Figure 4a, we found that at a smaller bubble size (20 Xe atoms), Xe atoms at the GB remain at the GB plane. However, with the growth of Xe bubbles, most Xe atoms in the larger bubbles are located in the bulk-like region away from the GB plane. This is attributed to the increase in temperature, which causes a change in the lattice constant and the reconstruction of the GB. An increase in temperature decreases the interfacial energy and it is easy to form segregation at the GB, where the energy of Xe atoms in the bulk is higher than that of the atoms at the GB. Thus, Xe atoms spontaneously converge toward the GB.

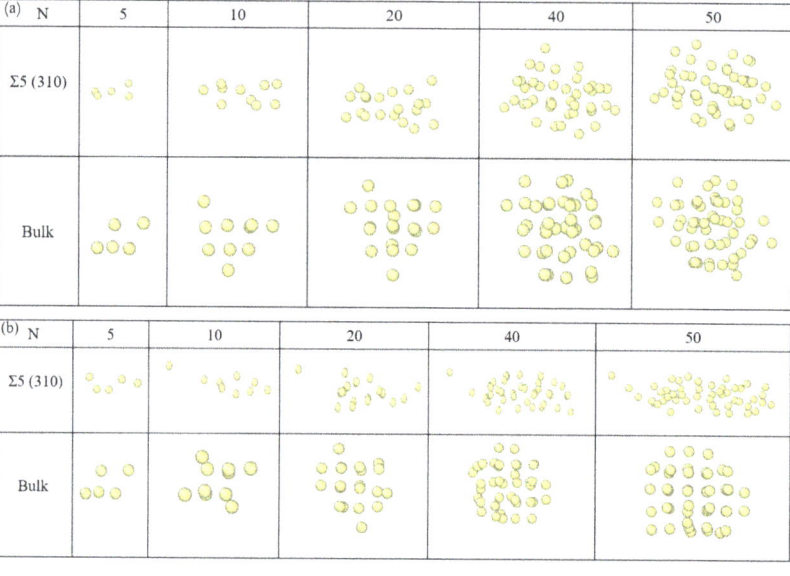

Figure 4. Comparison of the configurations bubbles at the Σ5(310) GB and bulk UO_2 at (**a**) 1000 K and (**b**) 2000 K.

Next, to analyze the characteristics of Xe bubbles, the volume (V) and pressure (P) of the bubbles after inserting each Xe atom were calculated. Figure 5 shows the relationship between the volume and pressure of Xe bubbles and the number of Xe atoms in the bubble. The inset of Figure 5a shows that the volume of Xe bubbles increases with an increase in temperature and, during bubble growth, the volume of Xe bubbles increases approximately linearly with an increase in the number of Xe atoms. However, at the same temperature, the volume of Xe bubbles at the GB is higher than that of the bubbles in the bulk. This is because GBs are surface defects in solid materials and there are many defects, such as vacancies, at the GBs. Thus, under the same conditions, the volume of Xe bubbles at the GB is larger than that of the bubbles in the bulk. Also, it reflects the influence of temperature on the volume of Xe bubbles, which is attributed to the increase in temperature, which changes the lattice constant and the reconstruction of GBs [19]. At 2000 K, the Σ5(310) GB consists of a more distorted triangular pattern, the middle gap is larger and the lattice constant increases. In Figure 5b, at the GB, the pressure of the Xe bubble is initially quite high and then drops with increasing Xe concentration until the number of Xe atoms reaches 10. When there are more Xe atoms in the bubbles, the pressure increases with an increase in the number of Xe atoms, which is consistent with that of Xe in the UO_2 bulk [4,8]. Then, with an increase in bubble growth, the bubble pressure generally decreases with some fluctuations involving several peaks and drops. This is because the bubbles may be far from the equilibrium, from which they grow because of rapid changes in temperature or insufficient local lattice vacancies [56]; thus, the smallest bubbles have a very high pressure, which results in a density comparable to that of solid Xe [57]. For larger bubbles, the density and pressure are relatively low. Experiments have shown that at the very early stages of bubble development, the bubble density is high with little evidence of deviation from circular bubbles. However, extensive bubble coalescence can greatly reduce the density of bubbles [58]. Under the same conditions, the pressure of Xe bubbles at the GB is less than that of the bubbles in the bulk. The internal pressure of Xe bubbles depends on not only the external stress on UO_2 but also the surface tension of the bubble voids [56].

In addition to volume and pressure, we evaluated the configuration of Xe bubbles at 1000 and 2000 K. Snapshots of the radial distribution function of Xe atoms at 1000 and 2000 K are shown in Figure 6. To avoid thermal fluctuations at 1000 and 2000 K, the positions of Xe atoms were obtained by taking an average total time of 50 ps before inserting the next Xe atom into the bubble. As shown in Figure 6, the peak is smaller and fuzzy at high temperatures because the atomic amplitude is large at high temperatures; thus, it is easy to deviate from the equilibrium position and approach the liquid state. With an increase in Xe atoms, Xe in the bubble evolves into a glassy/amorphous state, indicating that the behavior of Xe atoms in UO_2 is similar to that of a system with a hard spherical liquid in a solid. This observation is consistent with the results of Geng et al. [32]. For smaller bubbles (1–5 Xe atoms), the distribution of Xe atoms at the GB is more dispersed than that in the bulk, which also shows that Xe atoms at the GB diffuse more easily than those in the bulk, according to the MSD results. In the bulk, there is only one peak of Xe atoms at the beginning, indicating the uniform distribution of Xe atoms in the UO_2 matrix. With an increase in Xe atoms, the first peak gradually grows and the second peak appears, indicating the formation of Xe clusters and the emergence of the second layer of atoms in the clusters.

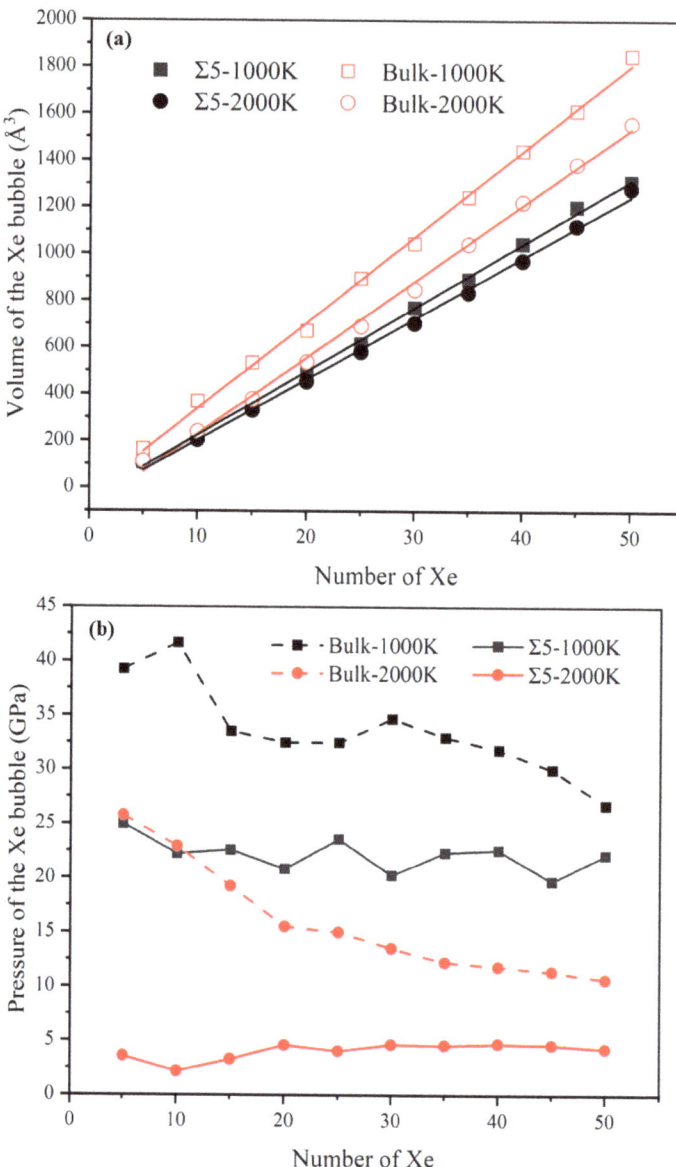

Figure 5. (**a**) Volume ($Å^3$) and (**b**) pressure (GPa) of Xe bubbles as a function of the number of Xe atoms in the bubbles. The lines are linear Arrhenius fits.

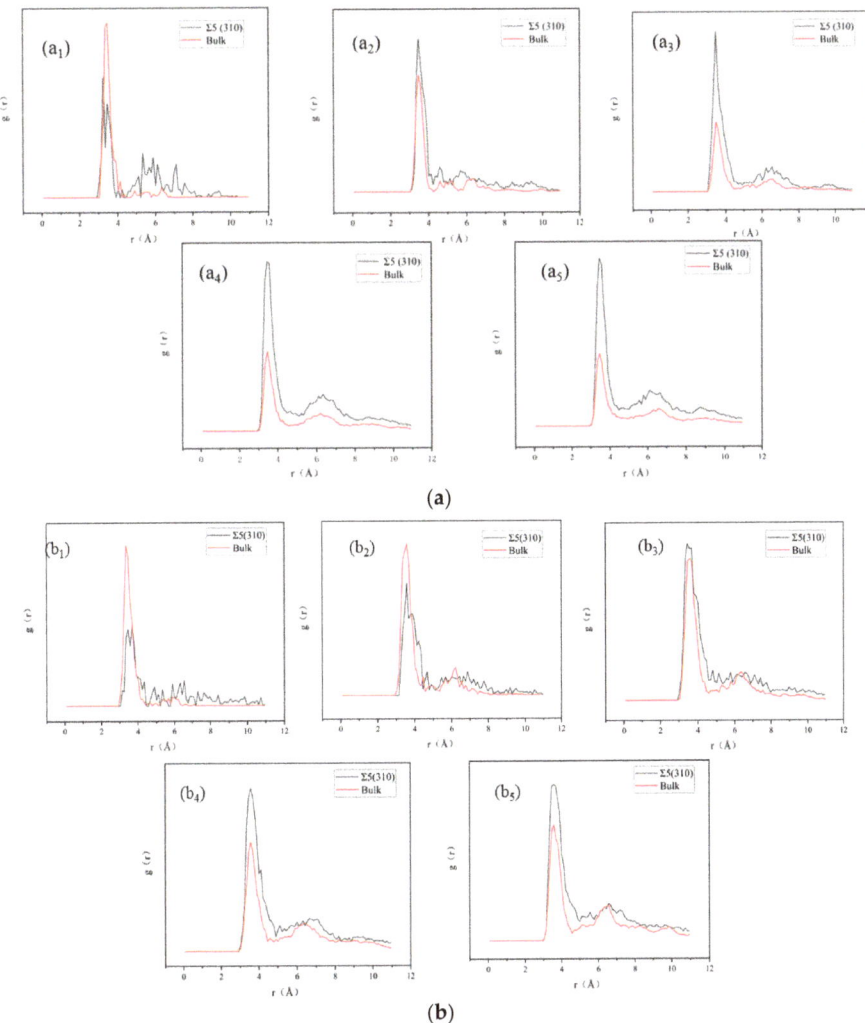

Figure 6. Radial distribution function of Xe atoms in Xe bubbles at (**a**) 1000 K and (**b**) 2000 K. The icons 1, 2, 3, 4 and 5 correspond to the number of Xe atoms 5, 10, 20, 40 and 50, respectively.

4. Conclusions

In this study, MD and static simulations were employed to investigate the energy and diffusion of a single Xe atom and small Xe clusters and the nucleation and growth of Xe bubbles at the UO_2 Σ5(310) GB. The results show that the formation energies of single Xe atoms and small Xe clusters at the GB are much lower than those in the bulk, indicating that impure Xe atoms are more stable at the GB and can precipitate into clusters. The diffusion activation energy of a single interstitial Xe atom at the GB is approximately 1 eV lower than that in the bulk. The energy barrier for interstitial Xe atoms to diffuse through the interstitial mechanism is high, which is mainly due to lattice deformation caused by the movement of Xe between two adjacent interstitial sites. The growth of Xe bubbles at the GB was simulated by sequentially inserting Xe atoms into a pre-existing Xe bubble. Then, the bubble volume and pressure were estimated for different numbers of Xe atoms. The pressure dropped with an increase in temperature, whereas the volume increased. At low

Xe concentrations, the Xe bubble volume increased approximately linearly with an increase in the number of Xe atoms. We also compared the configurations of Xe atoms at the GB and in the bulk at the same temperature and found that Xe atoms are more regular in the bulk, whereas, at the GB, multiple Xe atoms form a planar structure.

Author Contributions: Conceptualization, H.D.; Investigation, Y.X.; Methodology, Y.X., Z.W., L.W. (Li Wang), Y.C., Z.L., Q.W. and L.W. (Lu Wu); Validation, Y.X., Z.W., L.W. (Li Wang), Y.C., Z.L., Q.W. and L.W. (Lu Wu); Formal analysis, Y.X., L.W. (Li Wang), Y.C. and Z.L.; Resources, Z.W., Z.L., Q.W., L.W. (Lu Wu) and H.D.; Writing—original draft, Y.X.; Writing—review & editing, Y.X. and H.D.; Supervision, H.D. All authors have read and agreed to the published version of the manuscript.

Funding: This research received no external funding.

Institutional Review Board Statement: Not applicable.

Informed Consent Statement: Not applicable.

Data Availability Statement: Data sharing is not applicable to this article.

Conflicts of Interest: The authors declare no conflict of interest.

References

1. International Atomic Energy Agency. *Thermophysical Properties Database of Materials for Light Water Reactors and Heavy Water Reactors*; IAEA TECDOC Series No. 1496; International Atomic Energy Agency: Vienna, Austria, 2006.
2. Turnbull, J. The distribution of intragranular fission gas bubbles in UO_2 during irradiation. *J. Nucl. Mater.* **1971**, *38*, 203–212. [CrossRef]
3. Yang, L.; Wirth, B. Evolution of pressurized xenon bubble and response of uranium dioxide matrix: A molecular dynamics study. *J. Nucl. Mater.* **2021**, *544*, 152730. [CrossRef]
4. Xiao-Feng, T.; Chong-Sheng, L.; Zheng-He, Z.; Tao, G. Molecular dynamics simulation of collective behaviour of Xe in UO_2. *Chin. Phys. B* **2010**, *19*, 057102. [CrossRef]
5. Olander, D.R. *Fundamental Aspects of nuclear Reactor Fuel Elements*; TID-26711-Pl (Atomic Energy Commission, 1976); California University: Berkeley, CA, USA, 1976. [CrossRef]
6. Jeon, B.; Asta, M.; Valone, S.M.; Gronbech-Jensen, N. Simulation of ion-track ranges in uranium oxide. *Nucl. Instrum. Methods Phys. Res. Sect. B Beam Interact. Mater. At.* **2010**, *268*, 2688–2693. [CrossRef]
7. Govers, K.; Lemehov, S.; Verwerft, M. On the solution and migration of single Xe atoms in uranium dioxide–An interatomic potentials study. *J. Nucl. Mater.* **2010**, *405*, 252–260. [CrossRef]
8. Liu, X.-Y.; Andersson, D. Molecular dynamics study of fission gas bubble nucleation in UO_2. *J. Nucl. Mater.* **2015**, *462*, 8–14. [CrossRef]
9. Nerikar, P.V.; Rudman, K.; Desai, T.G.; Byler, D.; Unal, C.; McClellan, K.J.; Phillpot, S.R.; Sinnott, S.B.; Peralta, P.; Uberuaga, B.P. Grain boundaries in uranium dioxide: Scanning electron microscopy experiments and atomistic simulations. *J. Am. Ceram. Soc.* **2011**, *94*, 1893–1900. [CrossRef]
10. Ray, I.; Thiele, H. Transmission electron microscopy study of fission product behaviour in high burnup UO_2. *J. Nucl. Mater.* **1992**, *188*, 90–95. [CrossRef]
11. Djourelov, N.; Marchand, B.; Marinov, H.; Moncoffre, N.; Pipon, Y.; Nédélec, P.; Toulhoat, N.; Sillou, D. Variable energy positron beam study of Xe-implanted uranium oxide. *J. Nucl. Mater.* **2013**, *432*, 287–293. [CrossRef]
12. Sabathier, C.; Vincent, L.; Garcia, P.; Garrido, F.; Carlot, G.; Thome, L.; Martin, P.; Valot, C. In situ TEM study of temperature-induced fission product precipitation in UO_2. *Nucl. Instrum. Methods Phys. Res. Sect. B Beam Interact. Mater. At.* **2008**, *266*, 3027–3032. [CrossRef]
13. Niemiec, M.; Gadomski, A.; Łuczka, J.; Schimansky-Geier, L. Phase transformation kinetics in d-dimensional grains-containing systems: Diffusion-type model. *Phys. A Stat. Mech. Its Appl.* **1998**, *248*, 365–378. [CrossRef]
14. Gadomski, A. Kinetic Approach to the Nucleation-and-Growth Phase Transition in Complex Systems. *Nonlinear Phenom. Complex Syst. Minsk.* **2000**, *3*, 321–352.
15. Tschopp, M.A.; Horstemeyer, M.; Gao, F.; Sun, X.; Khaleel, M. Energetic driving force for preferential binding of self-interstitial atoms to Fe grain boundaries over vacancies. *Scr. Mater.* **2011**, *64*, 908–911. [CrossRef]
16. Lejcek, P. *Grain Boundary Segregation in Metals*; Springer Science & Business Media: Cham, Switzerland, 2010; Volume 136.
17. Van Brutzel, L.; Vincent-Aublant, E. Grain boundary influence on displacement cascades in UO2: A molecular dynamics study. *J. Nucl. Mater.* **2008**, *377*, 522–527. [CrossRef]
18. Zhang, Y.; Millett, P.C.; Tonks, M.R.; Bai, X.-M.; Biner, S.B. Molecular dynamics simulations of intergranular fracture in UO2 with nine empirical interatomic potentials. *J. Nucl. Mater.* **2014**, *452*, 296–303. [CrossRef]
19. Williams, N.R.; Molinari, M.; Parker, S.C.; Storr, M.T. Atomistic investigation of the structure and transport properties of tilt grain boundaries of UO_2. *J. Nucl. Mater.* **2015**, *458*, 45–55. [CrossRef]

20. Liu, L.; Chen, Y.; Gao, N.; Hu, W.; Xiao, S.; Gao, F.; Deng, H. Atomistic simulations of the interaction between transmutation-produced Re and grain boundaries in tungsten. *Comput. Mater. Sci.* **2020**, *173*, 109412. [CrossRef]
21. Gao, N.; Ghoniem, A.; Gao, X.; Luo, P.; Wei, K.; Wang, Z. Molecular dynamics simulation of Cu atoms interaction with symmetrical grain boundaries of BCC Fe. *J. Nucl. Mater.* **2014**, *444*, 200–205. [CrossRef]
22. Li, X.; Liu, W.; Xu, Y.; Liu, C.; Fang, Q.; Pan, B.; Chen, J.-L.; Luo, G.-N.; Wang, Z. An energetic and kinetic perspective of the grain-boundary role in healing radiation damage in tungsten. *Nucl. Fusion* **2013**, *53*, 123014. [CrossRef]
23. Plimpton, S. Fast parallel algorithms for short-range molecular dynamics. *J. Comput. Phys.* **1995**, *117*, 1–19. [CrossRef]
24. Govers, K.; Lemehov, S.; Hou, M.; Verwerft, M. Comparison of interatomic potentials for UO_2. Part I: Static calculations. *J. Nucl. Mater.* **2007**, *366*, 161–177. [CrossRef]
25. Cooper, M.; Rushton, M.; Grimes, R. A many-body potential approach to modelling the thermomechanical properties of actinide oxides. *J. Phys. Condens. Matter* **2014**, *26*, 105401. [CrossRef]
26. Jelea, A.; Pellenq, R.-M.; Ribeiro, F. An atomistic modeling of the xenon bubble behavior in the UO_2 matrix. *J. Nucl. Mater.* **2014**, *444*, 153–160. [CrossRef]
27. Thompson, A.E.; Meredig, B.; Wolverton, C. An improved interatomic potential for xenon in UO_2: A combined density functional theory/genetic algorithm approach. *J. Phys. Condens. Matter* **2014**, *26*, 105501. [CrossRef]
28. Brillant, G.; Gupta, F.; Pasturel, A. Fission products stability in uranium dioxide. *J. Nucl. Mater.* **2011**, *412*, 170–176. [CrossRef]
29. Chartier, A.; Van Brutzel, L.; Freyss, M. Atomistic study of stability of xenon nanoclusters in uranium oxide. *Phys. Rev. B* **2010**, *81*, 174111. [CrossRef]
30. Yun, Y.; Kim, H.; Kim, H.; Park, K. Atomic diffusion mechanism of Xe in UO_2. *J. Nucl. Mater.* **2008**, *378*, 40–44. [CrossRef]
31. Basak, C.; Sengupta, A.; Kamath, H. Classical molecular dynamics simulation of UO_2 to predict thermophysical properties. *J. Alloy. Compd.* **2003**, *360*, 210–216. [CrossRef]
32. Geng, H.; Chen, Y.; Kaneta, Y.; Kinoshita, M. Molecular dynamics study on planar clustering of xenon in UO_2. *J. Alloy. Compd.* **2008**, *457*, 465–471.
33. Morelon, N.-D.; Ghaleb, D.; Delaye, J.-M.; Van Brutzel, L. A new empirical potential for simulating the formation of defects and their mobility in uranium dioxide. *Philos. Mag.* **2003**, *83*, 1533–1555.
34. Cooper, M.; Kuganathan, N.; Burr, P.; Rushton, M.; Grimes, R.; Stanek, C.; Andersson, D. Development of Xe and Kr empirical potentials for CeO_2, ThO_2, UO_2 and PuO_2, combining DFT with high temperature MD. *J. Phys. Condens. Matter* **2016**, *28*, 405401. [CrossRef] [PubMed]
35. Scheiber, D.; Pippan, R.; Puschnig, P.; Romaner, L. Ab initio calculations of grain boundaries in bcc metals. *Model. Simul. Mater. Sci. Eng.* **2016**, *24*, 035013. [CrossRef]
36. Zheng, H.; Li, X.-G.; Tran, R.; Chen, C.; Horton, M.; Winston, D.; Persson, K.A.; Ong, S.P. Grain boundary properties of elemental metals. *Acta Mater.* **2020**, *186*, 40–49. [CrossRef]
37. Watanabe, T. Grain boundary engineering: Historical perspective and future prospects. *J. Mater. Sci.* **2011**, *46*, 4095–4115. [CrossRef]
38. Randle, V. Grain boundary engineering: An overview after 25 years. *Mater. Sci. Technol.* **2010**, *26*, 253–261. [CrossRef]
39. Lee, S.B.; Sigle, W.; Kurtz, W.; Rühle, M. Temperature dependence of faceting in Σ5 (310)[001] grain boundary of $SrTiO_3$. *Acta Mater.* **2003**, *51*, 975–981. [CrossRef]
40. He, H.; Ma, S.; Wang, S. Survey of Grain Boundary Energies in Tungsten and Beta-Titanium at High Temperature. *Materials* **2021**, *15*, 156. [CrossRef]
41. Borde, M.; Germain, A.; Bourasseau, E. Molecular dynamics study of UO_2 symmetric tilt grain boundaries around [001] axis. *J. Am. Ceram. Soc.* **2021**, *104*, 2879–2893. [CrossRef]
42. Galvin, C.O.; Cooper, M.W.D.; Fossati, P.; Stanek, C.R.; Grimes, R.W.; Andersson, D. Pipe and grain boundary diffusion of He in UO_2. *J. Phys. Condens. Matter* **2016**, *28*, 405002. [CrossRef]
43. Bourasseau, E.; Mouret, A.; Fantou, P.; Iltis, X.; Belin, R.C. Experimental and simulation study of grain boundaries in UO_2. *J. Nucl. Mater.* **2019**, *517*, 286–295. [CrossRef]
44. Stukowski, A. Visualization and analysis of atomistic simulation data with OVITO–the Open Visualization Tool. *Model. Simul. Mater. Sci. Eng.* **2009**, *18*, 015012. [CrossRef]
45. Dillon, S.J.; Harmer, M.P. Multiple grain boundary transitions in ceramics: A case study of alumina. *Acta Mater.* **2007**, *55*, 5247–5254. [CrossRef]
46. Sheppard, D.; Terrell, R.; Henkelman, G. Optimization methods for finding minimum energy paths. *J. Chem. Phys.* **2008**, *128*, 134106. [CrossRef]
47. Henkelman, G.; Uberuaga, B.P.; Jónsson, H. A climbing image nudged elastic band method for finding saddle points and minimum energy paths. *J. Chem. Phys.* **2000**, *113*, 9901–9904. [CrossRef]
48. Henkelman, G.; Jónsson, H. Improved tangent estimate in the nudged elastic band method for finding minimum energy paths and saddle points. *J. Chem. Phys.* **2000**, *113*, 9978–9985. [CrossRef]
49. Rycroft, C.H.; Grest, G.S.; Landry, J.W.; Bazant, M.Z. Analysis of granular flow in a pebble-bed nuclear reactor. *Phys. Rev. E* **2006**, *74*, 021306. [CrossRef]
50. Yang, L.; Gao, F.; Kurtz, R.J.; Zu, X.; Peng, S.; Long, X.; Zhou, X. Effects of local structure on helium bubble growth in bulk and at grain boundaries of bcc iron: A molecular dynamics study. *Acta Mater.* **2015**, *97*, 86–93. [CrossRef]

51. Chen, D.; Gao, F.; Deng, H.-Q.; Liu, B.; Hu, W.-Y.; Sun, X. Migration of defect clusters and xenon-vacancy clusters in uranium dioxide. *Int. J. Mod. Phys. B* **2014**, *28*, 1450120. [CrossRef]
52. Thompson, A.E.; Wolverton, C. First-principles study of noble gas impurities and defects in UO_2. *Phys. Rev. B* **2011**, *84*, 134111. [CrossRef]
53. Yun, Y.; Eriksson, O.; Oppeneer, P.M.; Kim, H.; Park, K. First-principles theory for helium and xenon diffusion in uranium dioxide. *J. Nucl. Mater.* **2009**, *385*, 364–367. [CrossRef]
54. Ichinomiya, T.; Uberuaga, B.P.; Sickafus, K.E.; Nishiura, Y.; Itakura, M.; Chen, Y.; Kaneta, Y.; Kinoshita, M. Temperature accelerated dynamics study of migration process of oxygen defects in UO_2. *J. Nucl. Mater.* **2009**, *384*, 315–321. [CrossRef]
55. Zhang, B.; Li, Y.-H.; Zhou, H.-B.; Deng, H.; Lu, G.-H. Segregation and aggregation of rhenium in tungsten grain boundary: Energetics, configurations and strengthening effects. *J. Nucl. Mater.* **2020**, *528*, 151867. [CrossRef]
56. Parfitt, D.C.; Grimes, R.W. Predicting the probability for fission gas resolution into uranium dioxide. *J. Nucl. Mater.* **2009**, *392*, 28–34. [CrossRef]
57. Nogita, K.; Une, K. High resolution TEM observation and density estimation of Xe bubbles in high burnup UO_2 fuels. *Nucl. Instrum. Methods Phys. Res. Sect. B Beam Interact. Mater. At.* **1998**, *141*, 481–486. [CrossRef]
58. White, R.J. The development of grain-face porosity in irradiated oxide fuel. *J. Nucl. Mater.* **2004**, *325*, 61–77. [CrossRef]

Article

Effect of Radiation Defects on Thermo–Mechanical Properties of UO₂ Investigated by Molecular Dynamics Method

Ziqiang Wang [1,2], Miaosen Yu [2], Chen Yang [1], Xuehao Long [2], Ning Gao [2,3,*], Zhongwen Yao [4,*], Limin Dong [5] and Xuelin Wang [2]

[1] Key Laboratory of Bionic Engineering, Ministry of Education, Jilin University, Changchun 130022, China; wangziqiang2021@gmail.com (Z.W.); chenyangjlu@126.com (C.Y.)
[2] Institute of Frontier and Interdisciplinary Science, Key Laboratory of Particle Physics and Particle Irradiation (MOE), Shandong University, Qingdao 266237, China; 202120966@mail.sdu.edu.cn (M.Y.); 17862969617@163.com (X.L.); xuelinwang@sdu.edu.cn (X.W.)
[3] Institute of Modern Physics, Chinese Academy of Sciences, Lanzhou 730000, China
[4] Department of Mechanical and Materials Engineering, Queen's University, Kingston, ON K7L 3N6, Canada
[5] School of Materials Science and Chemical Engineering, Harbin University of Science and Technology, Harbin 150080, China; donglimin@hrbust.edu.cn
* Correspondence: ning.gao@sdu.edu.cn (N.G.); zwyao@jlu.edu.cn (Z.Y.)

Abstract: Nuclear fuel performance is deteriorated due to radiation defects. Therefore, to investigate the effect of irradiation-induced defects on nuclear fuel properties is essential. In this work, the influence of radiation defects on the thermo-mechanical properties of UO₂ within 600–1500 K has been studied using the molecular dynamics method. Two types of point defects have been investigated in the present work: Frenkel pairs and antisites with concentrations of 0 to 5%. The results indicate that these point defects reduce the thermal expansion coefficient (α) at all studied temperatures. The elastic modulus at finite temperatures decreases linearly with the increase in concentration of Frenkel defects and antisites. The extent of reduction (R) in elastic modulus due to two different defects follows the trend $R_f > R_a$ for all studied defect concentrations. All these results indicate that Frenkel pairs and antisite defects could degrade the performance of UO₂ and should be seriously considered for estimation of radiation damage in nuclear fuels used in nuclear reactors.

Keywords: uranium dioxide; Frenkel pairs; antisites; molecular dynamics; elastic modulus; thermal expansion

Citation: Wang, Z.; Yu, M.; Yang, C.; Long, X.; Gao, N.; Yao, Z.; Dong, L.; Wang, X. Effect of Radiation Defects on Thermo–Mechanical Properties of UO₂ Investigated by Molecular Dynamics Method. *Metals* **2022**, *12*, 761. https://doi.org/10.3390/met12050761

Academic Editor: Enrique Jimenez-Melero

Received: 8 April 2022
Accepted: 25 April 2022
Published: 29 April 2022

Publisher's Note: MDPI stays neutral with regard to jurisdictional claims in published maps and institutional affiliations.

Copyright: © 2022 by the authors. Licensee MDPI, Basel, Switzerland. This article is an open access article distributed under the terms and conditions of the Creative Commons Attribution (CC BY) license (https://creativecommons.org/licenses/by/4.0/).

1. Introduction

Uranium dioxide (UO₂) is widely used as a nuclear fuel in the nuclear industry for various nuclear power reactors [1]. Thus, the safe operation of a nuclear reactor correlates strongly with the stability of UO₂. However, under extreme conditions, different radiation defects (e.g., vacancies, interstitials, and voids) would be created within the nuclear fuel due to irradiation. These defects would lead to severe degradation of the physical, thermal, and mechanical properties of the nuclear fuels [2–4]. For example, irradiation-induced fission products and vacancies can produce bubbles and voids, causing swelling and fragmentation which thus deteriorates the performance of fuels [5]. Therefore, to investigate the effect of radiation-induced defects on the thermo-mechanical properties of uranium dioxide is essential.

In the literature, numerous experimental and theoretical studies have been performed to understand the impact of fission products, porosities, and other defects on thermal transport in UO₂. For example, Hobson et al. analyzed porous UO₂ with porosity levels of 4.11 to 8.58% and observed the relationship of reductions in thermal conductivity to the temperature [6]. An experimental study on the effect of soluble fission products on thermal conductivity was also performed, which found that at lower temperatures the thermal conductivity would decrease with an increase in fission product concentration; however, at higher temperatures the concentration of fission products has only a slight influence on thermal transport [7].

In addition to experiments, computer simulations have also provided an effective tool to analyze the specific mechanism of fuels at the atomic level. Liu et al. investigated the effect of uranium vacancies, oxygen vacancies, and fission products on the thermal conductivity of uranium. They also observed a stronger effect on the reduction in thermal conductivity by uranium vacancies compared to that by oxygen vacancies [8]. Chen et al. performed molecular dynamics (MD) simulations to examine the effect of Xe bubble size and pressure on the thermal conductivity of uranium dioxide and demonstrated that the dispersed Xe atoms could result in a lower thermal conductivity than by clustering them into bubbles [9]. Uchida et al. performed molecular dynamics simulations to evaluate the effect of Schottky defects on the thermal properties in UO_2 and reported that thermal conductivity decreased with the increasing concentration of Schottky defects [10]. Furthermore, the thermal transport of ThO_2, as an alternative to conventional uranium nuclear fuel, was also investigated extensively. For example, Park et al. [11] investigated the effect of vacancies and substitutional defects on the thermal transport of ThO_2 by employing reverse non-equilibrium molecular dynamics (NEMD). The authors reported that the effect of thorium vacancy defects on the thermal transport of ThO_2 is even more detrimental than that of oxygen vacancy defects. In addition, compared to vacancy defects, substitutional defects in ThO_2 slightly affect the thermal transport [11]. To investigate the effect of irradiation-induced fission products on the thermal conductivity of thorium dioxide, Rahman et al. [5] examined the effect of Xe and Kr with impurity concentrations of 0 to 5% on the thermal conductivity of ThO_2 with the molecular dynamics method, and found that Xe and Kr resulted in significant reductions in the thermal conductivity of ThO_2.

In order to use nuclear fuels safely in reactors, the mechanical feature of fuels after irradiation is also an important property which needs to be considered [12]. For example, Jelea et al. examined the thermo-mechanical properties of a UO_2 matrix containing different concentrations of porosity and observed that the elastic modulus decreased with an increase in porosity concentration [13]. Rahman et al. examined the effect of fission products (Xe and Kr) and porosity on mechanical properties of ThO_2 within 300–1500 K using molecular dynamics simulations. By comparing the effect of fission products and porosity, the authors reported that the fission products resulted in a stronger reduction in elastic modulus than the porosity [14].

Although the effects of fission products and porosity on thermo-mechanical properties of UO_2 have been studied by different groups, to our best knowledge no investigation has been performed about the effect of Frenkel defects and antisites on the thermo-mechanical properties of irradiated UO_2. Considering its importance, in this work, the influences of Frenkel defects and antisites on the thermal expansion coefficient and elastic modulus of uranium dioxide are investigated extensively via molecular dynamics simulations. The thermal expansion coefficient of perfect and damaged systems is evaluated from changes in lattice parameters. Three independent elastic constants are calculated for each system, which are used to estimate the elastic modulus. The reduction in the elastic modulus induced by Frenkel defects and antisites is also calculated as a function of concentrations of defects in the system. A comparison is finally made between the effects of Frenkel defects and antisite defects to provide more understanding about the structure and property changes of UO_2 after irradiation. In the following sections, the computational method is first presented. The results and discussion are provided in Section 3. The conclusion is made in the last section.

2. Computational Method

In this work, MD simulations are performed using the LAMMPS (Large scale Atomic/ Molecular Massively Parallel Simulator) code (Sandia National Laboratories, Livermore, CA, USA) [15], which is a classical molecular dynamics (MD) code used to simulate the atomic interaction of selected materials. The interatomic potential developed by Cooper, Rushton, and Grimes (CRG) is used in this work for U-U, U-O and O-O interactions [4], which has been proven to reliably predict the various thermo-mechanical properties within the temperature range of 300 K to 3000 K [8,16–18]. In order to accurately describe the properties of UO_2, the Coulomb interaction is further included with the original pair. The computational box used in this work is a 10 × 10 × 12 extension of fluorite (CaF_2) unit cells containing 14,400 atoms. The lattice parameter for the computational box is the equilibrated lattice parameter at the investigated temperature. The periodic boundary condition is applied in all directions.

Generally, the O/U ratio in all defects after a displacement cascade in UO_2 is close to two [19], which is in agreement with the results presented by Devanathan et al. [20] and Van Brutzel et al. [21]. In order to create this structure of uranium dioxide with Frenkel defects and maintain the neutral charge of the system, uranium and oxygen atoms are removed from the system by keeping 1:2 ratio. The same amount of uranium and oxygen atoms are then randomly inserted at the octahedral interstitial positions of the face-centered cubic (fcc) cation sublattice [22]. Different from Frenkel defects, oxygen-antisites are created by substituting O atoms with U atoms. Similarly, uranium-antisites are created by substituting U atoms with O atoms. In order to maintain the stoichiometry of the defected system, the number of O-antisites is equal to that of U-antisites. In order to investigate the effect of defect concentration, UO_2 structures with 1%, 2%, and 5% Frenkel defects and 1%, 2%, and 5% antisite defects are built for further simulations. It should be noted that in the present work the concentration is defined as the value before MD relaxation at given temperatures. The main reason is that the relaxations at different temperatures could result in different concentration values after full relaxation. In order to avoid this misunderstanding during the investigation of concentration effects in this work, the concentration value before MD relaxation is used. For each defect concentration the statistical results are made based on 3 samples by randomly creating Frenkel or antisite defects.

The simulation box is first relaxed by the conjugated gradient (CG) method at 0 K and further relaxed for 500 ps at the temperature of interest under NPT (constant number, pressure, and temperature) ensemble with zero external pressure. It should also be noted the system containing 5% defects requires longer equilibration time (600 ps) to reach the equilibrium state. The equilibration is checked by the dependence of total energy and volume of the system on simulation time, which both indicate that the system has reached the equilibrium state before further calculations of lattice constants, thermal expansion coefficient, and elastic modulus. Therefore, the results obtained in the present work are reliable and could provide useful information to understand the properties of UO_2 after irradiation. The timestep of 1 fs is used for all simulation processes.

The lattice parameter as a function of temperature for different concentrations of Frenkel defects and antisite defects is first calculated. The thermal expansion coefficient (α) is then determined from the first derivate of the lattice parameter with respect to the temperature using the following Equation (1):

$$\alpha(T) = \frac{1}{L}\left(\frac{\partial L}{\partial T}\right)_p \quad (1)$$

where L is lattice parameter and $\left(\frac{\partial L}{\partial T}\right)_p$ is the slope of the plot for the lattice parameter as a function of temperature at the given temperature [23]. It should be noted that, in the present work, the structure of UO_2 is FCC and thus the thermal expansion coefficient is isotropic.

As the structure of uranium dioxide is cubic, three independent elastic constants (C_{11}, C_{12}, and C_{44}) need to be calculated. These elastic constants can be calculated by applying elementary strain in six directions and measuring the changes in the six stress components. In this work, the strain to induce the deformation of the simulation box was set to be 10^{-5}. Based on the dependence of the stress on the strain, these constants are calculated as described in [24]. Based on these three constants, the bulk modulus (B), shear modulus

(G), and Young's modulus (Y) can be calculated. The bulk modulus is calculated with the following equations [14].

$$B = (C_{11} + 2C_{12})/3 \tag{2}$$

Furthermore, according to the Hill's suggestion [25], in order to determine the shear modulus, the shear modulus (G_V) using the Voigt method and the shear modulus (G_R) using the Reuss method need to be obtained, which can be determined using the following Equations (3) and (4), respectively.

$$G_V = (C_{11} - C_{12} + 3C_{44})/5 \tag{3}$$

$$G_R = (5(C_{11} - C_{12})C_{44})/(4C_{44} + 3(C_{11} - C_{12})) \tag{4}$$

Thus, the shear modulus (G) using Hill's method can be obtained as the arithmetic average of G_V and G_R.

$$G = (G_V + G_R)/2 \tag{5}$$

Based on the calculated bulk modulus and shear modulus, Young's modulus is calculated with the following Equation (6).

$$Y = 9BG/(3B + G) \tag{6}$$

3. Results and Discussion

3.1. Effect of Defects on Lattice Parameter and Thermal Expansion Coefficient

The lattice parameter (L) of pure UO_2 as the function of temperature is plotted in Figure 1. The error bars in the figure correspond to the standard deviation calculated among the five different statistical lattice constants calculated at the given temperature. For comparison, the results from the VASP calculation by Wang et al. [26] and from the experimental measurement by Taylor et al. [27], Yamashita et al. [28], and Momin et al. [29] are also included in Figure 1. It is also clear that L increases linearly with increases in temperature from 0 K to 1500 K as investigated in this work. From these results, the results from the present MD agree better with the experimental value than those from VASP calculations. Thus, the MD method and the related empirical potential could be used for the present purpose for further simulations.

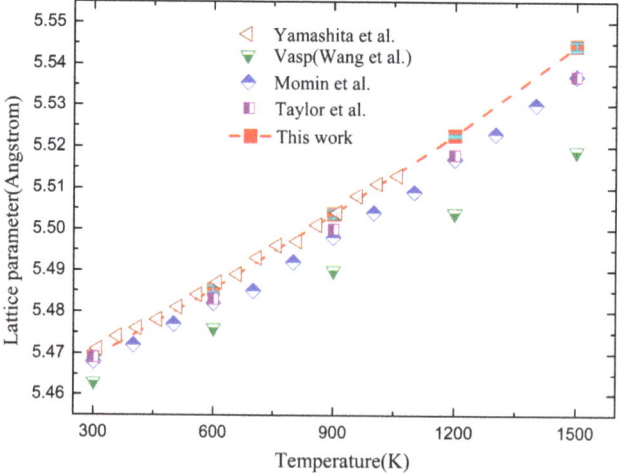

Figure 1. Dependence of lattice constant on temperature including the results from the present work, the experiment studies by Taylor et al. [27], Momin et al. [29], Yamashita et al. [28], and the VASP results by Wang et al. [26].

When defects are formed after irradiation, the effects of radiation defects are also simulated in this work. In Figure 2, the dependence of the lattice parameter of UO$_2$ on defect concentration at different temperatures is provided for Frenkel defects (dash) and antisite defects (solid). From Figure 2, it is clear that the lattice parameter of the system increases with an increase in Frenkel defect concentration from 0 to 5% at all investigated temperatures in the present work, although the increases are limited around 1.0%.

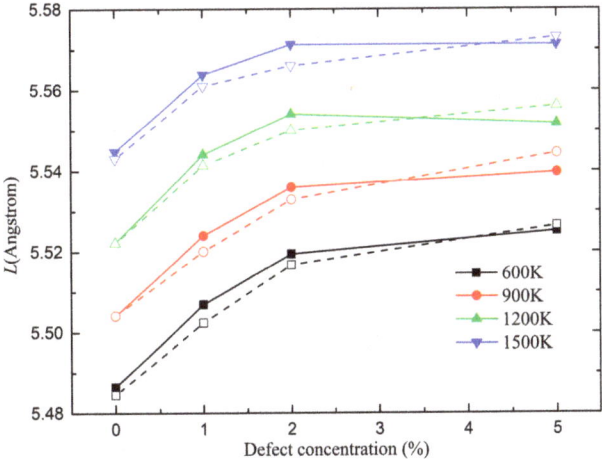

Figure 2. Dependence of lattice constant of UO$_2$ on defect concentration at different temperatures for Frenkel defects (dash) and antisite defects (solid).

The present results indicate that the formation of Frenkel pairs or antisite defects could increase the lattice constant quickly when the defect concentration is less than 2.0%, above which the lattice constant is almost a constant or increases or decreases slightly. Thus, the results are different from previous studies about the effect of porosity on the lattice parameter L of ThO$_2$, which reported that the L of the system increased linearly with increases in porosity concentration. The reason for the above difference may be mainly from the property of anisotropic effects induced by antisite defects and interstitials in Frenkel pairs, which is different from the isotropic vacancy or porosity. It should also be noted here that in this work, only defect concentrations up to 5% are considered. If higher concentrations were considered, the lattice constant may change accordingly.

As stated previously, because of the high temperature within the fuel due to the fission reaction, thermal expansion coefficient is considered to be an important factor for modelling and predicting the nuclear fuel's behavior [13]. Figure 3 presents the effects of Frenkel defects and antisite defects on the thermal expansion coefficient of UO$_2$ as a function of temperature. The uncertainty of the thermal expansion coefficient at different temperatures has also been calculated by changing the defect distribution but keeping the same concentration as stated in the computational method section. As shown in Figure 3, the uncertainty is limited for the three cases investigated in the present work. For comparison, the experimental results [27], MD derived values [30], and first principles data [31] for pure UO$_2$ are included in the figure.

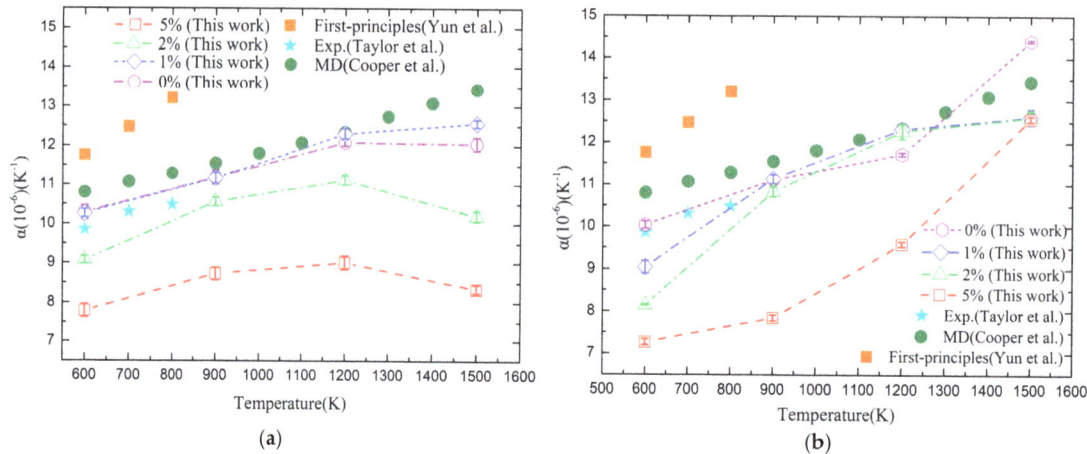

Figure 3. Thermal expansion coefficient for pure UO_2 and for defective UO_2 with different percentages of Frenkel defect (**a**), and antisites (**b**), as a function of temperature. For comparison, the experiment studies by Taylor et al. [27], the MD results by Cooper et al. [30] and the first-principles data by Yun et al. [31] are included in the figure.

Figure 3a shows that compared to the perfect system, the thermal expansion coefficient of systems containing 1% Frenkel defects has similar results to that of perfect systems at temperatures from 600 K to 1200 K, while at 1500 K the 1% Frenkel defects induce larger thermal expansion coefficients with an increase around 4.0%. When the system contains a higher concentration of Frenkel defects, e.g., 2.0% and 5.0%, the lower thermal expansion coefficients are observed. For example, when the concentration of Frenkel defects is 2%, the maximum reduction of thermal expansion coefficient up to 15% is observed at 1500 K. When Frenkel defect concentration is 5%, the maximum reduction is around 30% within 600–1500 K. Thus, it is found that the reduction of the thermal expansion coefficient increases with the increase in the concentration of Frenkel defects. Furthermore, it can also be observed from Figure 3a that when the defect concentration is low, e.g., 1%, the thermal expansion coefficient increases with an increase in temperature, while the higher concentration defects (2% and 5%) result in an increase in the thermal expansion coefficient for systems from 600 K to 1200 K, but decrease from 1200 K to 1500 K. In fact, according to Sun et al. [32], the thermal expansion coefficient can be described as a function of both temperature (T) and atomic restoring force of the system (F(r)), especially around the defect region. The F(r) is a function of atomic distance, which depends on the temperature and defect distribution in the computational box. Therefore, when the temperature changes, the change of F(r) together with temperature results in non-monotone temperature dependence of the thermal expansion coefficient.

Figure 3b presents the effect of antisite defects on the thermal expansion coefficient of UO_2. Different from the effects of Frenkel defects, Figure 3b clearly indicates that for the concentrations of antisites investigated in the present work, the thermal expansion coefficient of the system increases with an increase in temperature. When the temperature is 600 K, the thermal expansion coefficient decreases with an increase in antisite defect concentration. When the temperature is 900 K or 1200 K, the thermal expansion coefficient has similar values for systems containing antisite defects less than 2%, but decreases around 25–30% when the antisite defect concentration is 5%. When temperature is 1500 K, the thermal expansion coefficient has the same value for systems containing 1% to 5% antisite defects, which is lower than that of the perfect system. Comparing the results shown in Figure 3a,b, it could be found that antisite defects have stronger effects than Frenkel defects on the thermal expansion coefficient of UO_2.

3.2. Elastic Modulus of UO_2

After the calculation of elastic constants, the bulk modulus, Young's modulus, and shear modulus of perfect UO_2 are initially calculated from three independent elastic constants (C_{11}, C_{12}, and C_{44}) using Equations (2), (5) and (6). Figure 4 depicts the dependence of the bulk modulus of perfect UO_2 on temperature calculations of systems. For comparison, the experimental results from Belle et al. [33], the MD derived bulk modulus from Basak et al. [23], and ab initio data calculated by Wang et al. [26] are also included in this figure. It is clear that the present study has similar results to those obtained by previous MD calculations and experiments but lower than those from the VASP calculation. Figure 4 also indicates that the bulk modulus of perfect UO_2 derived in this study decreases with an increase in temperature, which has been confirmed in the previous study by Dorado et al. [34].

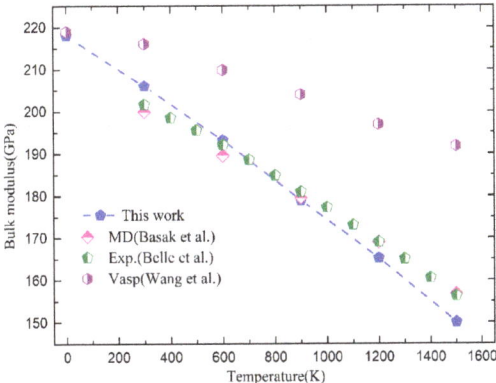

Figure 4. Variation of the bulk modulus of pure UO_2 versus temperature. For comparison, the experiment studies by Belle et al. [33], the Vasp data by Wang et al. [26] and the MD results by Basak et al. [23] are included.

Figure 5 shows the bulk modulus of the damaged UO_2 at different temperatures as a function of Frenkel defect concentration (dash) and antisite defect concentration (solid). Figure 5 shows that both a Frenkel defect and an antisite defect could considerably decrease the bulk modulus of UO_2, showing a linear decreasing dependence on temperature from 600 to 1500 K. With an increase in defect concentration, the elastic modulus also decreases accordingly, as shown by the figure. The extent of reduction in the elastic modulus for the system containing defects becomes smaller with increasing temperature and defect concentration. In addition, Figure 5 demonstrates that Frenkel defects increase the bulk modulus to a larger extent compared to that induced by antisite defects.

Figure 6 presents the effects of Frenkel (dash) and antisite defects (solid) on the shear modulus (G) of UO_2. Firstly, for the concentration of defects investigated in this work, the shear modulus decreases with an increase in temperature. The extent of reduction in the shear modulus decreases with increases in temperature from 600 to 1500 K. Similar to the effects on the bulk modulus shown in Figure 5, it can be seen from Figure 6 that the increase of defect concentration could significantly reduce the shear modulus of UO_2. However, the extent of reduction in G resulting from Frenkel defects is larger than that observed for antisite defects. For example, for 5% Frenkel and antisite defects there is a maximum of 20% and 17% reduction in G at all temperatures, respectively.

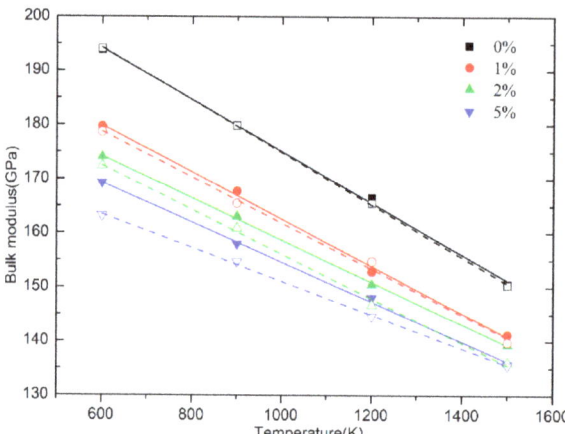

Figure 5. Variation of the bulk modulus of UO_2 containing different concentrations of Frenkel (dash) and antisite defects (solid) versus temperature. The fitted lines are also included in the figure.

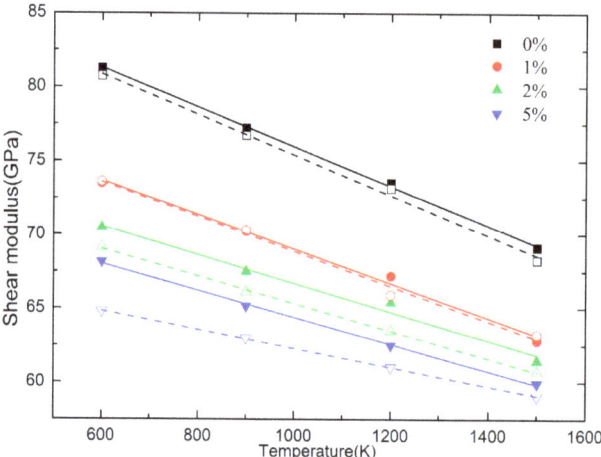

Figure 6. Variation of the shear modulus of UO_2 containing different concentrations of Frenkel (dash) and antisite (solid) defects versus temperature. The fitted lines are also included in the figure.

Young's modulus of UO_2 containing different concentrations of Frenkel (dash) and antisite defects (solid) as the function of the temperature is plotted in Figure 7. From this figure, it is clear that Young's modulus of uranium dioxide decreases linearly with the increase in temperature for Frenkel and antisite defects within the given concentrations. This result is similar to that reported by Jelea [13] et al. who observed that Young's modulus for damaged UO_2 with different percentages of porosity linearly decreases with increases in temperature. Similar to the results of the bulk modulus and the shear modulus, Young's modulus also decreases with increasing temperature and concentration of defects. The relative change in Y due to Frenkel defects is larger than that observed for antisite defects. Within the given temperatures, a maximum of 20% reduction in Y is observed for UO_2 systems containing 1%, 2%, and 5% Frenkel defects. In contrast, for antisite defects with the same concentrations, there is a maximum of 16% reduction in Y at all temperatures.

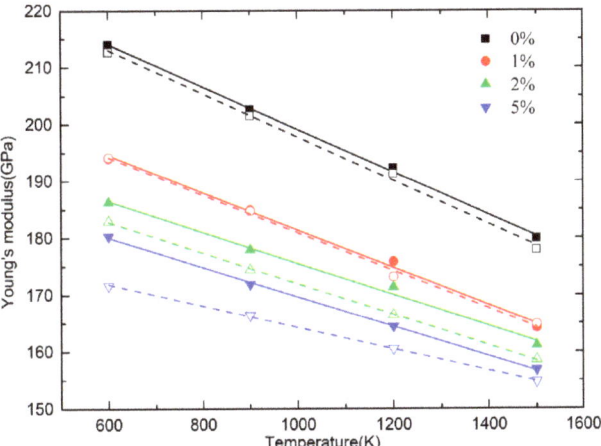

Figure 7. Variation of Young's modulus of UO$_2$ containing different concentrations of Frenkel (dash) and antisite (solid) defects versus temperature. The fitted lines are also included in the figure.

3.3. Reduction of Elastic Modulus of UO$_2$ by Frenkel Defects and Antisites

The effects of Frenkel defects and antisites on elastic modulus are reported in detail in Figures 8–10. In each of these figures, (a) and (b) represent the influence of Frenkel defects and antisite defects on the elastic modulus, respectively. The percentage of reduction (R) in the elastic modulus as the function of fractional point defects for all temperatures is plotted. The percentage of reduction of the elastic modulus is calculated using the following Equation (7).

$$R = (M_p - M)/M_p \tag{7}$$

where M_p and M represent the elastic modulus of a perfect and defective UO$_2$, respectively.

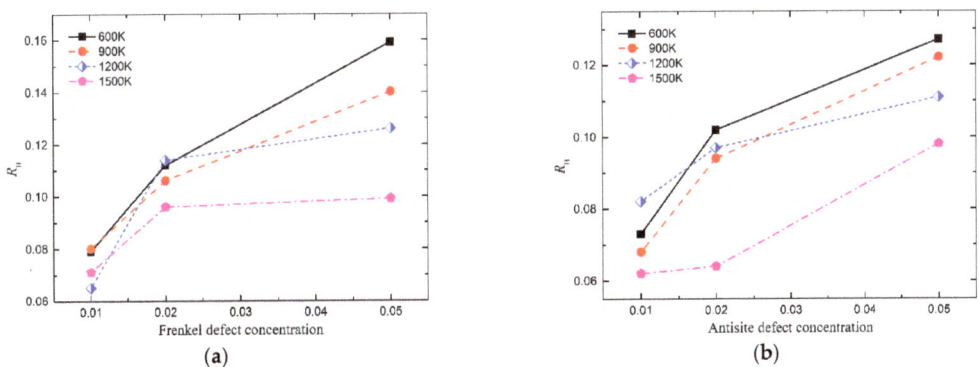

Figure 8. R_B as a function of the concentration of Frenkel defects (a) and antisites (b) at different temperatures.

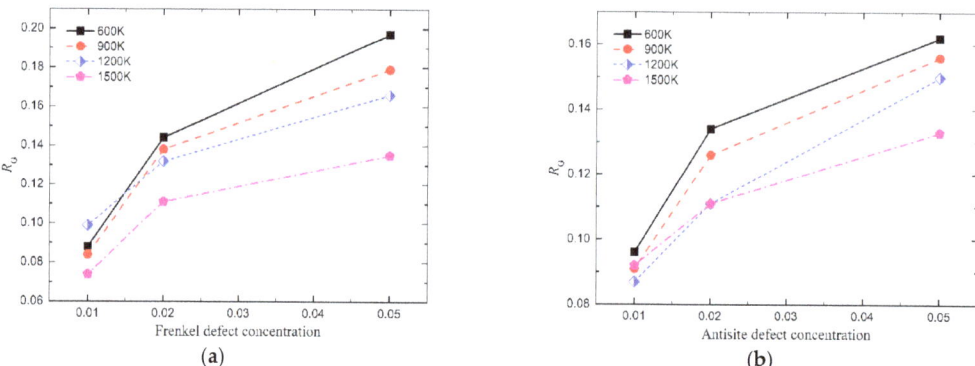

Figure 9. R_G as a function of the concentration of Frenkel defects (**a**) and antisites (**b**) at different temperatures.

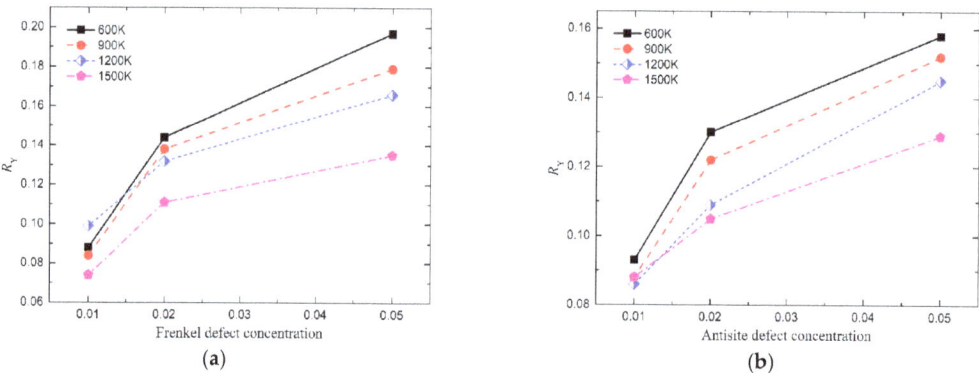

Figure 10. R_Y as a function of the concentration of Frenkel defects (**a**) and antisites (**b**) for different temperatures.

Figures 8a, 9a and 10a illustrate that in low concentration ranges the difference among the percentages of reduction in the elastic modulus for different temperatures is small. However, this difference is more evident at higher defect concentrations. In addition, this is not as significant for antisite defects as it is for Frenkel defects. The R of the elastic modulus due to antisite defects is smaller than that by Frenkel defects. For 5% defects, the degradation of the bulk modulus, shear modulus, and Young's modulus by Frenkel defects is higher than that of antisite defects by 3.4, 3.9, and 2.9%. Therefore, by comparing Figures 8–10, it can be observed that the percentage of reduction in the elastic modulus by Frenkel and antisite defects follows the trend $R_f > R_a$ for all studied defect concentrations.

4. Conclusions

In order to characterize the effect of irradiation on the performance of the nuclear fuel, it is necessary to investigate how the irradiation defects affect the thermal-mechanical property of nuclear fuel. In this study, the impact of Frenkel defects and antisites on thermal expansion and elastic constants has been examined in UO_2 via the molecular dynamics method in the temperature range of 600 to 1500 K. The results indicate that both Frenkel defects and antisite defects reduce the thermal expansion coefficient. However, the reduction in the thermal expansion coefficient due to antisite defects is larger than that observed for Frenkel defects. For the elastic modulus, the calculated bulk, shear, and Young's modulus of the pure UO_2 are in agreement with the experimental values.

Furthermore, the present results indicate that both Frenkel defects and antisite defects reduce the elastic modulus at all temperatures. The degree of reduction in the elastic modulus increases with increasing concentrations of defect. In addition, the percentage of reduction in the elastic modulus due to Frenkel and antisite defects follows the trend $R_f > R_a$ at all studied defect concentrations. All these calculated values can be used to predict the performance of UO_2 under irradiation used in the nuclear reactor environment.

Author Contributions: Conceptualization, Z.W. and N.G.; methodology, Z.W.; software, M.Y.; validation, C.Y., X.L. and Z.W.; formal analysis, Z.W.; investigation, M.Y.; resources, N.G.; data curation, Z.W.; writing—original draft preparation, Z.W.; writing—review and editing, N.G.; visualization, M.Y. and L.D.; supervision, N.G. and Z.Y.; project administration, Z.Y.; funding acquisition, N.G. and X.W. All authors have read and agreed to the published version of the manuscript.

Funding: This work was financially supported by the National Natural Science Foundation of China (Project Nos. 12075141 and 12175125).

Institutional Review Board Statement: Not applicable.

Informed Consent Statement: Not applicable.

Data Availability Statement: The data that support the findings of this study are available from the corresponding author upon reasonable request.

Conflicts of Interest: The authors declare no conflict of interest.

References

1. Cooper, M.W.D.; Murphy, S.T.; Rushton, M.J.D.; Grimes, R.W. Thermophysical properties and oxygen transport in the $(U_x,Pu_{1-x})O_2$ lattice. *J. Nucl. Mater.* **2015**, *461*, 206–214. [CrossRef]
2. Devanathan, R.; Van Brutzel, L.; Chartier, A.; Guéneau, C.; Mattsson, A.E.; Tikare, V.; Bartel, T.; Besmann, T.; Stan, M.; Van Uffelen, P. Modeling and simulation of nuclear fuel materials. *Energy Environ. Sci.* **2010**, *3*, 1406–1426. [CrossRef]
3. Martin, G.; Maillard, S.; Brutzel, L.V.; Garcia, P.; Dorado, B.; Valot, C. A molecular dynamics study of radiation induced diffusion in uranium dioxide. *J. Nucl. Mater.* **2009**, *385*, 351–357. [CrossRef]
4. Cooper, M.W.D.; Rushton, M.J.D.; Grimes, R.W. A many-body potential approach to modelling the thermomechanical properties of actinide oxides. *J. Phys. Condens. Matter* **2014**, *26*, 105401. [CrossRef] [PubMed]
5. Rahman, M.; Szpunar, B.; Szpunar, J. Dependence of thermal conductivity on fission-product defects and vacancy concentration in thorium dioxide. *J. Nucl. Mater.* **2020**, *532*, 152050. [CrossRef]
6. Hobson, I.C.; Taylor, R.; Ainscough, J.B. Effect of porosity and stoichiometry on the thermal conductivity of uranium dioxide. *J. Phys. D Appl. Phys.* **1974**, *7*, 1003–1015. [CrossRef]
7. Ishimoto, S.; Hirai, M.; Ito, K.; Korei, Y. Effects of Soluble Fission Products on Thermal Conductivities of Nuclear Fuel Pellets. *J. Nucl. Sci. Technol.* **1994**, *31*, 796–802. [CrossRef]
8. Liu, X.Y.; Cooper, M.W.D.; McClellan, K.J.; Lashley, J.C.; Byler, D.D.; Bell, B.D.C.; Grimes, R.W.; Stanek, C.R.; Andersson, D.A. Molecular Dynamics Simulation of Thermal Transport in UO_2 Containing Uranium, Oxygen, and Fission-product Defects. *Phys. Rev. Appl.* **2016**, *6*, 044015. [CrossRef]
9. Chen, W.; Cooper, M.W.D.; Xiao, Z.; Andersson, D.A.; Bai, X.-M. Effect of Xe bubble size and pressure on the thermal conductivity of UO_2—A molecular dynamics study. *J. Mater. Res.* **2019**, *34*, 2295–2305. [CrossRef]
10. Uchida, T.; Sunaoshi, T.; Kato, M.; Konashi, K. Thermal properties of UO_2 by molecular dynamics simulation. In *Progress in Nuclear Science and Technology*; Atomic Energy Society of Japan: Tokyo, Japan, 2011; Volume 2, pp. 598–602.
11. Park, J.; Farfán, E.B.; Mitchell, K.; Resnick, A.; Enriquez, C.; Yee, T. Sensitivity of thermal transport in thorium dioxide to defects. *J. Nucl. Mater.* **2018**, *504*, 198–205. [CrossRef]
12. Rahman, M.J.; Szpunar, B.; Szpunar, J.A. Comparison of thermomechanical properties of $(U_x,Th_{1-x})O_2$, $(U_x,Pu_{1-x})O_2$ and $(Pu_x,Th_{1-x})O_2$ systems. *J. Nucl. Mater.* **2019**, *513*, 8–15. [CrossRef]
13. Jelea, A.; Colbert, M.; Ribeiro, F.; Tréglia, G.; Pellenq, R.J.M. An atomistic modelling of the porosity impact on UO_2 matrix macroscopic properties. *J. Nucl. Mater.* **2011**, *415*, 210–216. [CrossRef]
14. Rahman, M.J.; Szpunar, B.; Szpunar, J.A. Effect of fission generated defects and porosity on thermo-mechanical properties of thorium dioxide. *J. Nucl. Mater.* **2018**, *510*, 19–26. [CrossRef]
15. Thompson, A.P.; Aktulga, H.M.; Berger, R.; Bolintineanu, D.S.; Brown, W.M.; Crozier, P.S.; in 't Veld, P.J.; Kohlmeyer, A.; Moore, S.G.; Nguyen, T.D.; et al. LAMMPS—A flexible simulation tool for particle-based materials modeling at the atomic, meso, and continuum scales. *Comput. Phys. Commun.* **2022**, *271*, 108171. [CrossRef]
16. Galvin, C.O.T.; Cooper, M.W.D.; Rushton, M.J.D.; Grimes, R.W. Thermophysical properties and oxygen transport in $(Th_x,Pu_{1-x})O_2$. *Sci. Rep.* **2016**, *6*, 36024. [CrossRef]

17. Cooper, M.; Middleburgh, S.; Grimes, R. Modelling the thermal conductivity of $(U_xTh_{1-x})O_2$ and $(U_xPu_{1-x})O_2$. *J. Nucl. Mater.* **2015**, *466*, 29–35. [CrossRef]
18. Qin, M.; Cooper, M.; Kuo, E.; Rushton, M.; Grimes, R.; Lumpkin, G.; Middleburgh, S. Thermal conductivity and energetic recoils in UO_2 using a many-body potential model. *J. Phys. Condens. Matter* **2014**, *26*, 495401. [CrossRef]
19. Martin, G.; Garcia, P.; Sabathier, C.; Van Brutzel, L.; Dorado, B.; Garrido, F.; Maillard, S. Irradiation-induced heterogeneous nucleation in uranium dioxide. *Phys. Lett. A* **2010**, *374*, 3038–3041. [CrossRef]
20. Devanathan, R.; Yu, J.; Weber, W.J. Energetic recoils in UO_2 simulated using five different potentials. *J. Chem. Phys.* **2009**, *130*, 174502. [CrossRef]
21. Van Brutzel, L.; Rarivomanantsoa, M. Molecular dynamics simulation study of primary damage in UO_2 produced by cascade overlaps. *J. Nucl. Mater.* **2006**, *358*, 209–216. [CrossRef]
22. Van Brutzel, L.; Chartier, A.; Crocombette, J.P. Basic mechanisms of Frenkel pair recombinations in UO_2 fluorite structure calculated by molecular dynamics simulations. *Phys. Rev. B* **2008**, *78*, 024111. [CrossRef]
23. Basak, C.B.; Sengupta, A.K.; Kamath, H.S. Classical molecular dynamics simulation of UO_2 to predict thermophysical properties. *J. Alloys Compd.* **2003**, *360*, 210–216. [CrossRef]
24. Dai, H.; Yu, M.; Dong, Y.; Setyawan, W.; Gao, N.; Wang, X. Effect of Cr and Al on Elastic Constants of FeCrAl Alloys Investigated by Molecular Dynamics Method. *Metals* **2022**, *12*, 558. [CrossRef]
25. Zuo, L.; Humbert, M.; Esling, C. Elastic properties of polycrystals in the Voigt-Reuss-Hill approximation. *J. Appl. Crystallogr.* **1992**, *25*, 751–755. [CrossRef]
26. Wang, B.-T.; Zhang, P.; Lizárraga, R.; Di Marco, I.; Eriksson, O. Phonon spectrum, thermodynamic properties, and pressure-temperature phase diagram of uranium dioxide. *Phys. Rev. B* **2013**, *88*, 104107. [CrossRef]
27. Taylor, D. Thermal expansion data. II: Binary oxides with the fluorite and rutile structures, $MO2$, and the antifluorite structure, $M2O$. *Trans. J. Br. Ceram. Soc.* **1984**, *83*, 32–37.
28. Yamashita, T.; Nitani, N.; Tsuji, T.; Inagaki, H. Thermal expansions of NpO_2 and some other actinide dioxides. *J. Nucl. Mater.* **1997**, *245*, 72–78. [CrossRef]
29. Momin, A.C.; Mirza, E.B.; Mathews, M.D. High temperature X-ray diffractometric studies on the lattice thermal expansion behaviour of UO_2, ThO_2 and $(U0.2Th0.8)O_2$ doped with fission product oxides. *J. Nucl. Mater.* **1991**, *185*, 308–310. [CrossRef]
30. Cooper, M.W.D.; Murphy, S.T.; Fossati, P.C.M.; Rushton, M.J.D.; Grimes, R.W. Thermophysical and anion diffusion properties of $(U_x, Th_{1-x})O_2$. *Proc. R. Soc. A Math. Phys. Eng. Sci.* **2014**, *470*, 20140427.
31. Yun, Y.; Legut, D.; Oppeneer, P.M. Phonon spectrum, thermal expansion and heat capacity of UO_2 from first-principles. *J. Nucl. Mater.* **2012**, *426*, 109–114. [CrossRef]
32. Sun, C.Q. An approach to local band average for the temperature dependence of lattice thermal expansion. *arXiv* **2008**, arXiv:0801.0771.
33. Belle, J.; Berman, R. *Thorium Dioxide: Properties and Nuclear Applications*; USDOE Assistant Secretary for Nuclear Energy: Washington, DC, USA, 1984.
34. Dorado, B.; Freyss, M.; Martin, G. GGA+U study of the incorporation of iodine in uranium dioxide. *Eur. Phys. J. B* **2009**, *69*, 203–209. [CrossRef]

Article

Effect of Cr and Al on Elastic Constants of FeCrAl Alloys Investigated by Molecular Dynamics Method

Hui Dai [1], Miaosen Yu [1], Yibin Dong [1], Wahyu Setyawan [2], Ning Gao [1,3,*] and Xuelin Wang [1]

[1] Institute of Frontier and Interdisciplinary Science and Key Laboratory of Particle Physics and Particle Irradiation (MOE), Shandong University, Qingdao 266237, China; huidai@mail.sdu.edu.cn (H.D.); 202120966@mail.sdu.edu.cn (M.Y.); ybdong@mail.sdu.edu.cn (Y.D.); xuelinwang@sdu.edu.cn (X.W.)
[2] Pacific Northwest National Laboratory, Richland, WA 99352, USA; wahyu.setyawan@pnnl.gov
[3] Institute of Modern Physics, Chinese Academy of Sciences, Lanzhou 730000, China
* Correspondence: ning.gao@sdu.edu.cn

Abstract: The FeCrAl alloy system is recognized as one of the candidate materials for accident-tolerant fuel (ATF) cladding in the nuclear power industry due to its high oxidation resistance under irradiation and high-temperature environments. The concentrations of Cr and Al have a significant effect on elastic properties of the FeCrAl alloy. In this work, elastic constants C_{11}, C_{12}, C_{44}, bulk modulus and shear modulus of FeCrAl alloy were calculated with molecular dynamics methods. We explored compositions with 1–15 wt.% Cr and 1–5 wt.% Al at temperatures from 0 K to 750 K. The results show that the concentrations of Al and Cr have different effects on the elastic constants. When the concentration of Al was fixed, a decrease in bulk modulus and shear modulus with increasing Cr content was observed, consistent with previous experimental results. The dependence of elastic constants on temperature was also the same as in the experiments. Investigations into elastic properties of defect-containing alloys have shown that vacancies, voids, interstitials and Cr-rich precipitations have different effects on elastic properties of FeCrAl alloys. Investigations of elastic properties of defect-containing alloys have shown that vacancies, void, interstitials and Cr-rich precipitations have different effects on elastic properties of FeCrAl alloys. Therefore, the present results indicate that both the Cr and Al concentrations and radiation defects should be considered to develop and apply the FeCrAl alloy in ATF design.

Keywords: FeCrAl; elastic properties; temperature effect; molecular dynamics

Citation: Dai, H.; Yu, M.; Dong, Y.; Setyawan, W.; Gao, N.; Wang, X. Effect of Cr and Al on Elastic Constants of FeCrAl Alloys Investigated by Molecular Dynamics Method. *Metals* **2022**, *12*, 558. https://doi.org/10.3390/met12040558

Academic Editor: Angelo Fernando Padilha

Received: 18 January 2022
Accepted: 22 March 2022
Published: 25 March 2022

Publisher's Note: MDPI stays neutral with regard to jurisdictional claims in published maps and institutional affiliations.

Copyright: © 2022 by the authors. Licensee MDPI, Basel, Switzerland. This article is an open access article distributed under the terms and conditions of the Creative Commons Attribution (CC BY) license (https://creativecommons.org/licenses/by/4.0/).

1. Introduction

After the Fukushima Daiichi nuclear power disaster in 2011, development of accident-tolerant fuels (ATF) has been in great demand to enhance the safety of nuclear power plants, especially under accident conditions [1]. In addition to coated claddings, FeCrAl cladding is one of the possible technologies for use in ATF. In the past few years, based on this concept, FeCrAl alloys have been extensively studied and the advantages have been recognized, including improved high-temperature steam oxidation resistance [2], resistance to radiation damages [3], strength under both normal conditions and high-temperature accident conditions [4], and enhanced corrosion performance under normal conditions [5].

The roles of Cr and Al elements have been explored to understand the above properties. For example, Cr is added to improve the water corrosion resistance of these alloys at normal operating temperatures, and Al is added to improve the high-temperature oxidation resistance of the alloy through the formation of protective α-Al_2O_3 [6,7]. However, it should also be noted that various properties of FeCrAl alloys are strongly affected by the addition and concentrations of Cr and Al, including the mechanical properties. For example, the α' phase in FeCrAl alloy will nucleate and grow at temperatures around 475 °C with a concentration of Cr, C_{Cr}, around 10~18 at.%, which leads to embrittlement of the alloy under normal conditions [8,9]. Meanwhile, under irradiation, the α' phase could form even at 320 °C, with C_{Cr} around 10~18 at.% [10]. Hence, it is necessary to decrease the concentration of Cr to reduce the α' phase formation and related embrittlement of the alloy under irradiation at medium temperatures [11]. For this purpose, Oak Ridge National Laboratory (ORNL) has prepared a variety of FeCrAl-Y alloys with different Cr concentrations to develop new low-Cr, high-strength FeCrAl-Y alloys [11]. They found that a FeCrAl alloy with C_{Cr} less than 13 wt.% can effectively avoid the formation of a brittle phase at 475 °C. Gussev et al. found that adding Al to the alloy can improve the oxidation resistance, but the higher Al concentration, C_{Al}, increases, the greater the difficulty of welding and manufacturing the FeCrAl alloys [8,12].

In addition, previous studies have also reported that the concentration of Cr and Al could significantly affect the mechanical properties of FeCrAl alloys [12]. The elastic properties have been investigated for evaluating the survivability of cladding materials under energetic neutron irradiation [13]. In fact, mechanical properties, including the elastic modulus, yield stress, ultimate tensile stress, and so on, have been generally measured and estimated for selection of structural materials used in reactors before neutron irradiation experiments [14,15]. For example, before irradiation, it was reported by Liu et al. that the increase in Cr concentration will enhance the bulk modulus of the alloy. However, when the concentration of Al increases from 4.5 wt.% to 6 wt.%, the bulk modulus could also slightly decrease [14]. From experimental viewpoints, Speich et al. measured the shear modulus G, the ratio of the bulk modulus to the shear modulus K/G, and the Poisson's ratio v [16] using an ultrasonic pulse echo repeat method [17] at 298 K, indicating that the shear modulus and Poisson's ratio of the Fe-rich region of the Fe-Cr alloy showed a linear dependence with the increase of Cr concentration. Harmouche et al. prepared FeAl-B2 phase samples with Fe atomic fractions from 50.87% to 60.2%, and by measuring its Young's modulus at room temperature, the authors found that the dependence of Young's modulus on the concentration of Fe shows a quadratic relationship [18]. The dependence of elastic modulus, Poisson's ratio, and shear modulus on temperature has also been explored for FeCrAl alloy and almost no variation has been observed as a function of major alloy elements [12]. Based on the above literature review, it is clear that although there are many studies on the mechanical properties of binary alloys, research on the elastic properties of ternary alloys needs to be enriched, especially on the mutual effect of Cr and Al on the elastic constants of a ternary FeCrAl alloy. From the viewpoint of fundamental research, it would be necessary to investigate the Cr and Al concentration effect from the lower to higher values in order to present the possible dependence of elastic properties on these alloy elements.

Different from the effect of Cr and Al on mechanical properties before the irradiation, the effect of radiation-induced defects on the mechanical properties of FeCrAl alloy is another key factor in evaluating the performance of the selected material. From previous studies, it is well known that the elastic modulus is one of important parameters to be dynamically measured in order to estimate the state of structure materials during the performance of nuclear reactors [13]. Thus, understanding the elastic constant and modulus before and after the irradiation is important for the development and further optimization of FeCrAl alloys used in ATF design. Under irradiation, the super-saturated interstitials, vacancies, voids, and precipitates are generally observed, which are expected to affect the mechanical properties of FeCrAl alloys. For example, the formation of Cr-rich a' phases in irradiated FeCrAl alloys has been investigated by small-angle neutron scattering (SANS) [19,20], atom probe tomography (APT) [21,22], high-efficiency STEM-EDS (energy-dispersive spectroscopy) [21] and a series of technical studies. Unfortunately, although the irradiation hardening and embrittlement of FeCrAl after irradiation have been investigated, the detailed dependence of elastic constants or modulus on irradiation defects in FeCrAl alloys is not well understood, so research comparing this to the dynamically measured elastic modulus of materials to increase the safe operation of nuclear reactors would be valuable.

Therefore, in order to shed light on above questions, in this work, molecular dynamics (MD) simulations have been performed to calculate the elastic constants of FeCrAl alloy with 1~15 wt.% Cr and 1~5 wt.% Al at different temperatures (0 K, 300 K, 450 K, 600 K and 750 K) to understand the concentration effect of alloy elements on elastic properties of FeCrAl alloy. The effects of radiation defects, including vacancies, interstitials and Cr precipitates on the elastic properties, are also simulated. Through these calculations, the elastic properties of FeCrAl alloy before and after the irradiation can be understood from an atomic scale. In the rest of this paper, the method of calculations is introduced in Section 2, results and discussion are provided in Section 3, and conclusion is given in the last section.

2. Methods

In this work, molecular dynamics (MD) simulations have been performed with LAMMPS software to calculate the elastic constants at different temperatures. The computational box was built with three directions along [100], [010] and [001]. There are 10 unit cells along each direction, thus, 2000 atoms in the matrix box. The interactions between atoms in FeCrAl are described by the F-S potential developed by Liao et al., which has been shown to be accurate to describe the characteristics of the elements in FeCrAl alloys [22]. The time step is 1 femtosecond (fs), and the total simulation time is up to 10 ps for each simulation.

The elastic tensor can be written as a 6 × 6 matrix using the Voigt concept, that is, C_{ij} instead of C_{ijkl}. According to this definition, during calculations, the elastic constants can be calculated by applying elementary deformations in the six distinct strain components and measuring the changes in the six stress components accordingly. In this work, the strain to induce the deformation of the simulation box was set as 10^{-6}. Thus, according to the dependence of stress on strain, the elastic constants can be calculated. For temperatures higher than 0 K, thermal motion is included in the calculation of the stress. For the cubic lattice simulated in this work, the elastic tensor has three independent elastic constants, namely C_{11}, C_{12} and C_{44}. With the above method, these elastic constants can be calculated with or without temperature effect. In fact, this method has been applied by different groups to investigate the elastic properties of materials. For example, Miller et al. studied the elastic constants of PbTe, SnTe and $Ge_{0.08}Sn_{0.92}Te$ with this method [23]. In order to understand the influence of solute atoms on the elastic constants of FeCrAl alloys, 1~15 wt.% Cr and 1~5 wt.% Al atoms were introduced as substitutional solutes to form a solid solution, since FeCrAl alloys generally form a solid solution, as observed in previous experiments [10,24]. For convivence, the atomic concentration, at.%, is usually applied for atomic simulations. The conversion between wt.% and at.% can be obtained by including the atomic mass for Fe, Cr and Al. Examples of this unit conversion are listed in Table 1. It should also be noted that both Cr and Al atoms were introduced into the computational box in a randomly substitutional way by keeping the distance between each pair of substitutional atoms at least one lattice constant. As stated above, simulations at different temperatures (0 K, 300 K, 450 K, 600 K, 750 K) were performed to explore the effect of temperature on the elastic constants of the alloy. Furthermore, at each temperature, 50 simulations were performed to get reliable statistics and to estimate uncertainties in the results.

Table 1. Examples of conversion from wt.% to at.% for several FeCrAl alloys.

Solute	Fe1Cr1Al			Fe2Cr5Al			Fe5Cr3Al			Fe8Cr5Al			Fe13Cr5Al		
Element	Fe	Cr	Al	Fe	Cr	Al	Fe	Cr	Al	Fe	Cr	Al	Fe	Cr	Al
wt.%	98	1	1	93	2	5	92	5	3	87	8	5	82	13	5
at.%	97	1	2	88	2	10	89	5	6	82	8	10	77	13	10

In addition, in order to consider the effect of vacancies, interstitials and Cr precipitates on elastic constants, a big simulation model with 36,000 atoms was built with three directions in the same orientation as before. The radiation-induced vacancies (and voids) are included in the computational box by changing, n, the number of vacancies in each vacancy cluster or void and, m, the number of vacancy clusters or voids. For convenience, in this work, the total number of vacancies in the box was kept as 320, and thus $nm = 320$. The value of m is defined as 1, 2, 4, 5, 10, 20, 40, 80, 160 and 320, thus n is calculated accordingly. To investigate the effect of interstitials, different numbers of separated interstitials are were in the computational box. For a single interstitial, the <110> Fe-Fe or Fe-Cr dumbbell was built after the construction of FeCrAl alloy since the substitutional Al has lower formation energy, and thus, only the Fe-Fe and Fe-Cr dumbbells were built in the present work. As to the effect of Cr-rich α' on elastic properties, similar to the model of vacancy clusters, the precipitates containing different numbers of Cr atoms were considered. The number of precipitates was 1 to 7 and two cases with a total number Cr atoms in the box up to 4300 (C_{Cr} ~ 11%) and 4850 (C_{Cr} ~ 13%), respectively, were considered. Following above method, simulations at 0 K were performed to explore the effects of vacancies, interstitials and Cr precipitates on the elastic constants of the alloys. In each case, 50 simulations were also performed to obtain reliable statistics and estimate the uncertainty in the results.

3. Results and Discussion

3.1. Dependence of Elastic Constants on Concentration of Cr and Al

The calculated elastic constants, C_{11}, C_{12}, C_{44} and related bulk modulus and shear modulus, of FeCrAl alloy as a function of concentration of Cr and Al at 0 K are shown in Figure 1. The statistical uncertainties of these elastic constants and modulus were also calculated, which showed a similar dependence on concentration of Cr and Al, and thus just one example is presented in Figure 1f. Each panel in Figure 1 has five curves, corresponding to the five different concentrations of Al from 1 wt.% to 5 wt.%. For each curve, the data are plotted as a function Cr concentration from 1 wt.% to 15 wt.%. Some curves show an increasing trend, while others exhibit an opposite trend, indicating different dependencies of elastic constants on the concentrations of Al and Cr.

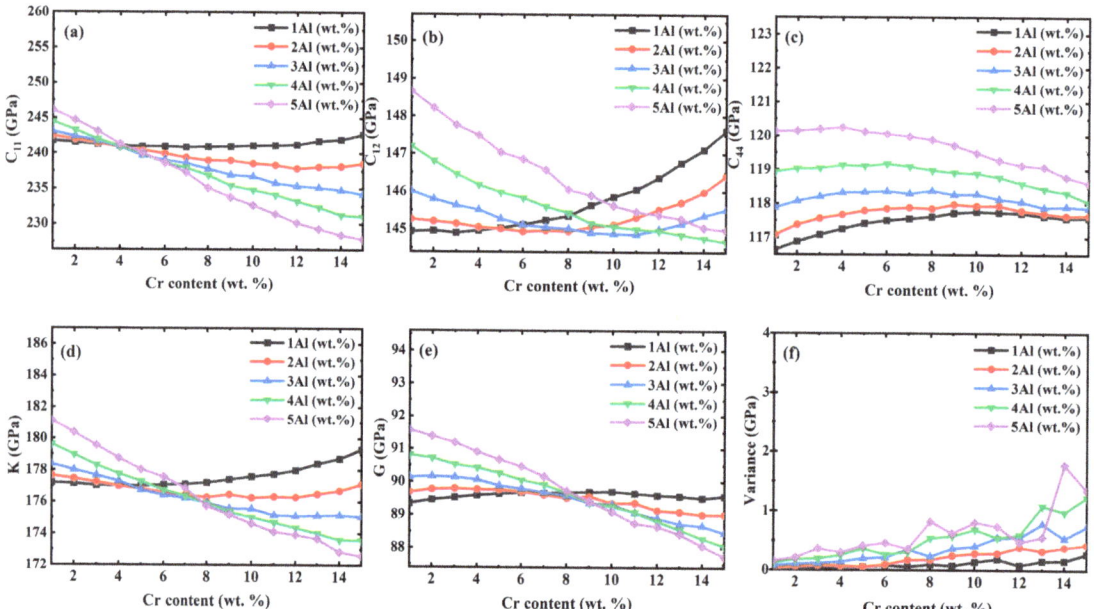

Figure 1. (a–e) Dependence of elastic constants, C_{11}, C_{12}, C_{44}, bulk modulus and shear modulus on Cr and Al concentrations at 0 K. (f) Example of statistical uncertainties for these elastic constants based on 50 simulations.

As shown in Figure 1a, for the lowest concentration of Al at 1 wt.%, C_{11} increases slightly with Cr concentration up to 15 wt.%. While for 2 to 5 wt.% Al, C_{11} decreases accordingly. Furthermore, from Figure 1a, when C_{Cr} is in the range of 1–3 wt.%, C_{11} increases with the increase of C_{Al}, while the opposite is true for C_{Cr} in the range of 5–15 wt.%. Interestingly, all curves intersect at a common point at 4 wt.% Cr, indicating that at this Cr concentration, C_{11} does not depend on C_{Al}.

For C_{12}, as shown in Figure 1b, when C_{Al} is 1 wt.% or 2 wt.%, C_{12} increases with C_{Cr} from 1 to 15 wt.%. When C_{Al} is 3 wt.%, C_{12} decreases firstly and then increases with C_{Cr}. When C_{Al} is 4 wt.% or 5 wt.%, C_{12} decreases with C_{Cr}. Furthermore, when C_{Cr} is fixed and C_{Cr} is lower than 5 wt.%, C_{12} increases with the increase of C_{Al}. When C_{Cr} is in the range of 5–13 wt.%, there is no clear dependence on C_{Al} with a fixed C_{Cr}. For C_{44}, as shown in Figure 1c, when C_{Al} is fixed from 1 to 5 wt.%, C_{44} increases firstly to a peak value and then decreases with the increase of C_{Cr}. It should also be noted that the peak is a function of C_{Al}, that is, the higher C_{Al} the higher the peak, and that the peak occurs at a smaller C_{Cr}. When C_{Cr} is fixed, C_{44} increases with the increase of C_{Al}.

In addition, the bulk modulus (K) and the shear modulus (G) have also been calculated according to the relationship between the elastic constants and K and G, derived by Voigt [25], as shown in Figure 1d,e. It is clear that when C_{Al} is 1 wt.%, the bulk modulus increases with C_{Cr}, which is consistent with the results of previous studies [26], that is, the bulk modulus of the alloy increases with the increase of Cr content. When C_{Al} is 2 wt.%, the bulk modulus decreases firstly and then increases with C_{Cr}. When C_{Al} is higher than 2 wt.%, the bulk modulus decreases with the increase of C_{Cr}. When the concentration of Cr is fixed, the bulk modulus increases with the increase of C_{Al} for C_{Cr} < 4 wt.%. For C_{Cr} in the range of 4–8 wt.%, the dependence of the bulk modulus on C_{Al} is not linear. For C_{Cr} > 8 wt.%, the bulk modulus decreases with the increase of C_{Al}. Regarding the shear modulus G, at 1 wt.% Al, the shear modulus is almost a constant with C_{Cr}. When the concentration of Al is higher than 1 wt.%, the shear modulus decreases with the increase of C_{Cr}. From this figure, there is also a common intersection of the shear modulus curves at a critical Cr concentration (8 wt.%), similar to the case of C_{11} (at 4 wt.%). Thus, for C_{Cr} < 8 wt.%, the shear modulus increases with Al content, while the opposite is true for C_{Cr} > 8 wt.%.

3.2. Dependence of Elastic Constants on Temperature

The elastic constants and modulus calculated at 300 K are shown in Figure 2. As shown in Figure 2a, the dependence of C_{11} on C_{Cr} and C_{Al} is similar to the results obtained at 0 K, as shown in Figure 1a. For C_{12}, comparing to the results shown in Figure 1b, for C_{Al} < 4 wt.%, a similar dependence of C_{12} on C_{Cr} can be observed, that is, C_{12} increases with C_{Cr}. However, for C_{Al} equal to 4 wt.% and 5 wt.%, different trends were obtained. C_{12} is almost a constant as a function of C_{Cr} for C_{Al} = 4 wt.%. While for C_{Al} = 5 wt.%, C_{12} decreases firstly with C_{Cr} up to around 7 wt.% and then remains almost constant, as shown in the Figure 2 b. Furthermore, for a fixed C_{Cr}, C_{12} always increases with the increase of C_{Al}. For C_{44}, C_{44} increases with the increase of both C_{Cr} and C_{Al}, a behavior that is different from the C_{44} at 0 K, as previously shown in Figure 1c. Compared to the results at 0 K, the bulk modulus shows a similar dependence on the concentration of Cr and Al, as shown in Figures 1d and 2d. The shear modulus also shows a similar trend, but the critical C_{Cr} concentration for G (the intersection point) has changed from ~8 to ~15 wt.%. From these results, especially from the bulk modulus and shear modulus, it can be estimated that from the mutual effects of Cr and Al concentration on elastic properties, the concentration of Al should be not higher than 3% and the related Cr concentration should be higher than 10% in the bulk FeCrAl alloys. However, it should also be noted that the above results were obtained only according to the effect of concentration of Cr and Al on elastic properties but without considering the other defects and related defect interactions. Furthermore, since the formation of Al_2O_3 on the material surface is the main reason to increase the high-temperature oxidation resistance of FeCrAl alloy and addition of Cr is necessary to increase the corrosion resistance, the Cr and Al concentrations are expected to be higher to satisfy this requirement. In fact, many researchers have suggested to add the other elements to increase the performance of FeCrAl alloy. Therefore, the optimization of Cr and Al concentration needs to consider various conditions, which is beyond the topic of the present paper. The optimization of FeCrAl alloy can be found elsewhere [26,27].

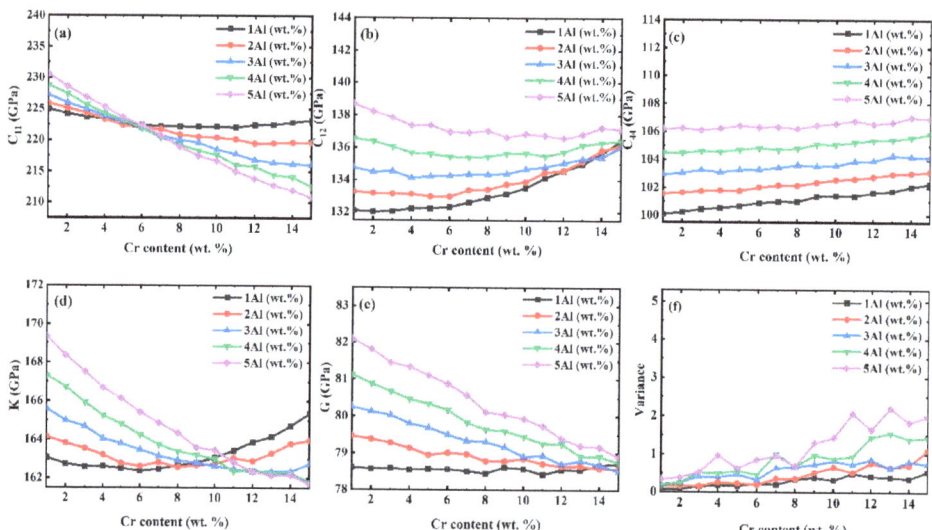

Figure 2. (**a**–**e**) Dependence of elastic constants, C_{11}, C_{12}, C_{44}, bulk modulus and shear modulus on Cr and Al concentration at 300 k. (**f**) Example of statistical uncertainty for these elastic constants based on 50 simulations.

In this work, as stated in the Methods section, the elastic constants and modulus at 450 K, 600 K and 750 K were also calculated. Detailed analysis indicates that the results obtained at different temperatures, including 0 K, 300 K, 450 K, 600 K and 750 K, show a similar dependence on the concentration of Cr and Al. One example for C_{Al} = 1 wt.% is shown in Figure 3, which clearly indicates the same dependence of elastic constants on Cr and Al concentration at different temperatures. Therefore, based on these results, the dependence of elastic constants and modulus on temperature can be explored.

The dependence of elastic properties on temperature is shown in Figure 4. The results for alloys with 1 wt.% Al and 4 wt.%, 8 wt.% and 14 wt.% Cr, are shown to illustrate the temperature effect. From this figure, it is evident that all elastic constants investigated in this work, C_{11}, C_{12}, C_{44}, bulk modulus and shear modulus of these three alloys, decrease with increasing temperature from 300 K to 750 K, which is also consistent with the results of a previous study [27]. It is worth noting that in a previous experiment [28], the decrease of elastic constants with the increase of temperature has also been observed, in which the bulk modulus decreased from about 200 GPa at 300 K to about 150 GPa at 750 K, and the shear modulus decreased from about 77 GPa at 300 K to about 65 GPa at 750 K, both of which decreased by about 15% [28]. In the present work, the bulk and shear moduli decrease by about 6% and 9%, respectively, less than in the experiment. The main reason is that in the experiment, the sample was polycrystalline, while a single crystal is employed in our calculations. In a polycrystalline material, the effect of grain boundaries could contribute to the decrease of the elastic properties.

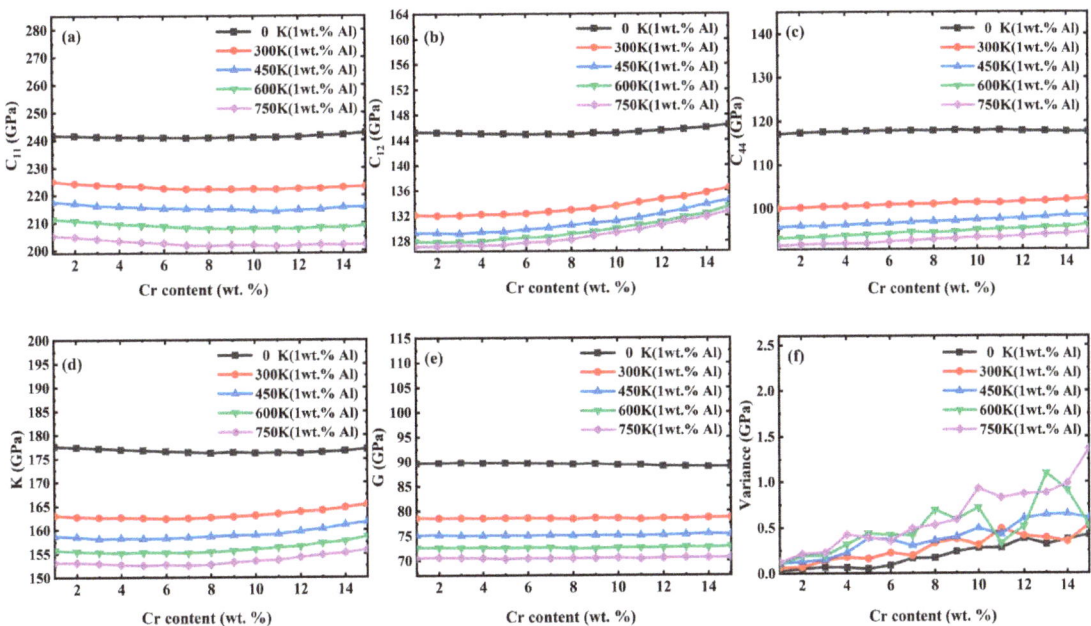

Figure 3. (a–e) Dependence of elastic constants, C_{11}, C_{12}, C_{44}, bulk modulus and shear modulus on the concentration of Cr in an FeCrAl alloy with 1 wt.% Al at temperatures 0~750 k, (f) example of statistical uncertainty for these elastic constants based on 50 simulations.

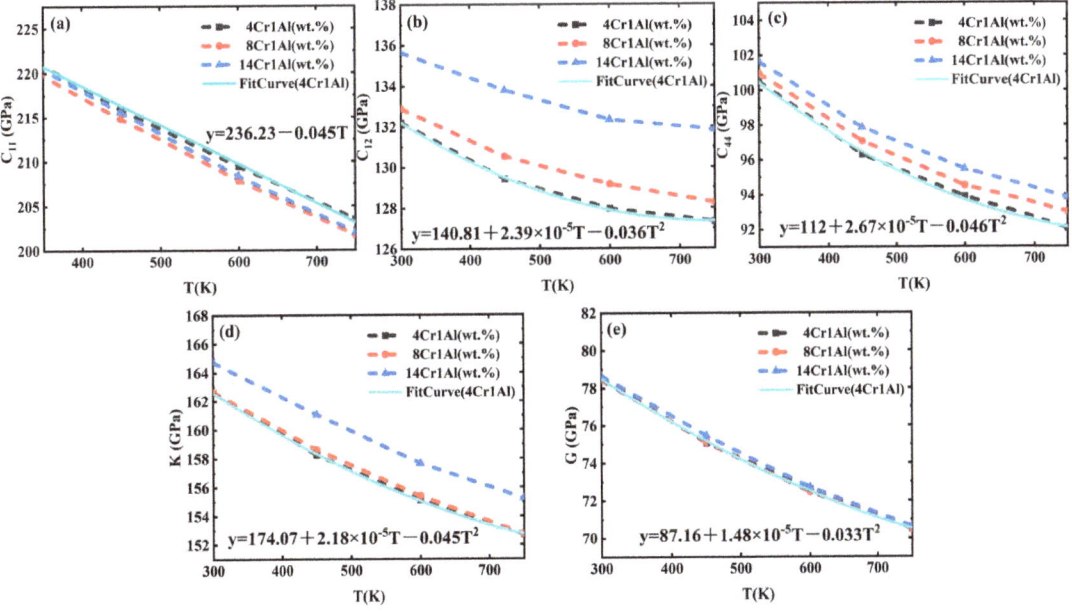

Figure 4. (a–e) Dependence of elastic constants, C_{11}, C_{12}, C_{44}, bulk modulus and shear modulus on temperature for 4Cr1Al (wt.%), 8Cr1Al (wt.%) and 14Cr1Al (wt.%) compositions at 300–750 K.

As shown in Figure 4, the dependence of each elastic constant on temperature follows the same trend for different concentration of Cr and Al, which, thus, can be described by the same equation but with different parameter values. For C_{11}, this dependence can be described as:

$$C_{11}(T) = A - BT \tag{1}$$

where A and B are parameters that depend on the concentration Cr and Al. For example, for 4Cr1Al (wt.%), 8Cr1Al (wt.%) and 14Cr1Al (wt.%), A is 236.23, 235.63 and 236.65 GPa, respectively, and B is 0.045, 0.045 and 0.047 GPa/K, respectively. The dependence of the other elastic constants, C_{12}, C_{44}, K and G, on temperature can be described by the following equation:

$$C(T) = A + BT - CT^2 \tag{2}$$

The fitted values for A, B and C for different concentrations of Cr and Al are listed in Table 2. In Figure 4, the fitted equations for FeCrAl alloy with $C_{Cr} \sim 4\%$(wt.%) and $C_{Al} \sim 1\%$(wt.%) are also included.

Table 2. Fitted parameter values in the models of elastic constants and modulus as a function of temperature for Fe4Cr1Al (wt.%), Fe8Cr1Al (wt.%) and Fe14Cr1Al (wt.%). The models are given in Equations (1) and (2).

Solute Constants	Fe4Cr1Al(wt.%)			Fe8Cr1Al(wt.%)			Fe14Cr1Al(wt.%)		
	A(GPa)	B(GPa/K)	C(GPa/K^2)	A(GPa)	B(GPa/K)	C(GPa/K^2)	A(GPa)	B(GPa/K)	C(GPa/K^2)
C_{11}	236.23	0.045	/	235.63	0.045	/	236.65	0.047	/
C_{12}	140.81	2.39×10^{-5}	0.036	139.61	1.64×10^{-5}	0.027	141.60	1.49×10^{-5}	0.024
C_{44}	112	2.67×10^{-5}	0.047	112.25	2.67×10^{-5}	0.046	112.11	2.37×10^{-5}	0.042
K	174.07	2.18×10^{-5}	0.045	172.68	1.52×10^{-5}	0.038	174.18	1.35×10^{-5}	0.035
G	87.16	1.48×10^{-5}	0.033	87.20	1.54×10^{-5}	0.034	86.90	1.26×10^{-5}	0.031

The nature of the decrease in the elastic constant after the temperature rises is due to the change in the lattice constant caused by thermal expansion [13]. This phenomenon can also be understood from the elastic constant temperature dependence formula derived by Leibfried and Ludwig in 1961 [29].

3.3. Influence of Radiation Defects on Elastic Properties

The effects of different vacancy types on elastic properties are shown in Figure 5. According to experimental results, the concentrations of Cr and Al in FeCrAl are generally from 12 wt.% and 5% respectively. In this work, the Fe8Cr5Al and Fe13Cr5Al alloys were selected as examples to show the effects of radiation defects on elastic constants. From these figures, it is clear that after the formation of radiation-induced vacancies, the C_{11}, C_{12}, C_{44}, bulk modulus and shear modulus of these two alloys decrease in comparison to the perfect state. It should also be noted that when the vacancy clusters are formed, these elastic constants decrease slightly or are kept constants and then increase with an increase of vacancy cluster size but decrease of the number of vacancy clusters. The K/G value remains around 2.30, larger than 1.75, which is still a ductility alloy according to Pugh's rule [30].

The dependence of elastic properties on different numbers of dumbbell interstitials are shown in Figure 6 for Fe8Cr5Al and Fe13Cr5Al alloys. It is also clear that C_{11}, C_{44} and shear modulus decrease when the interstitials are formed in these two alloys. However, C_{12} and bulk modulus increase when the number of interstitials increases, different from the results of C_{11}, C_{44} and shear modulus. Therefore, according to the definitions of bulk and shear modulus, it can be understood from the present results that the formation of interstitials could increase the ability of the alloys to resist the volume change under the effect of pressure but decrease the ability to resist the torsion applied on alloys. When the materials are applied in a nuclear reactor, the interaction between dislocations and these interstitials could further affect the mechanical properties. Thus, the present results indicate that the formation of an interstitial should be included, even if these interstitials are distributed separately and before their interaction with dislocations.

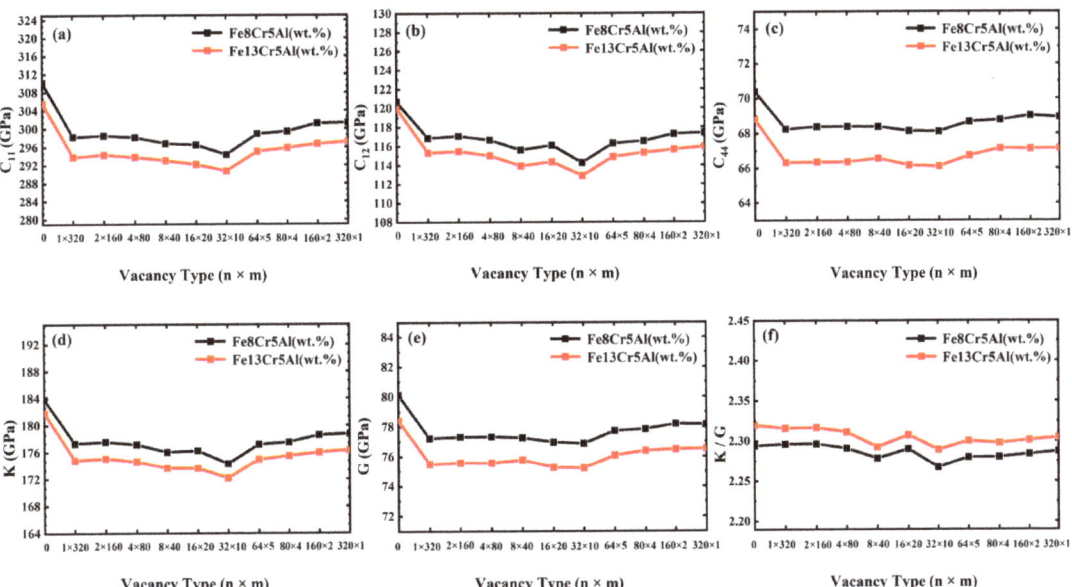

Figure 5. (a–e) The effects of vacancy and void on C_{11}, C_{12}, C_{44}, bulk modulus and shear modulus, (f) shows the value of K/G as a function of vacancy type. In these figures, n is the number of vacancies in each vacancy cluster or void and m is the number of vacancy clusters or voids.

When the Cr-rich precipitates are formed, the dependence of elastic properties on numbers of Cr precipitates are shown in Figure 7 for Fe8Cr5Al and Fe13Cr5Al alloys. From these results, it can be seen that after the Cr-rich precipitates are formed, these elastic properties of alloys increase but are kept as almost a constant, thus they are independent of the number and volume of precipitates. From previous studies, it is known that the precipitates would affect the mechanical properties. According to the present results, it can be understood that the effect of α′ precipitates on the mechanical properties of FeCrAl should be mainly due to their interaction with dislocations, different from the effects of vacancies, voids and interstitials, as shown in Figures 5 and 6. All results shown in Figures 5 and 6 indicate that an increase of Cr concentration would result in the decrease of elastic properties under the irradiation environment. Thus, the concentration of Cr in alloys should also be limited as indicated experimentally [3,9]. These results also indicate the radiation defects could result in larger decreases or increases of elastic properties than the Cr and Al concentrations. Furthermore, comparing the effects from vacancies, voids, interstitials and Cr-rich precipitates, it is clear that the interstitials have stronger effects on bulk and shear modulus than vacancies, voids and Cr-rich precipitates. It should also be noted the present results only consider the effects from radiation defects. The effect from the interaction between dislocations and radiation defects on mechanical properties should be investigated in future work.

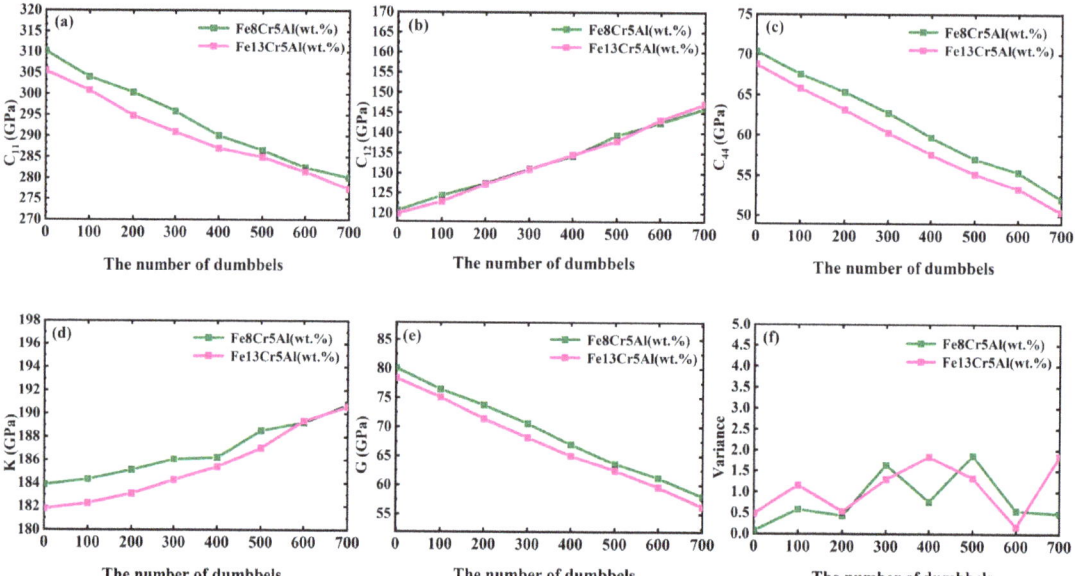

Figure 6. (a–e) The dependence of C_{11}, C_{12}, C_{44}, bulk modulus and shear modulus on number of dumbbell interstitials, (f) shows one example of statistical uncertainty for these elastic constants based on 50 simulations at 0 K.

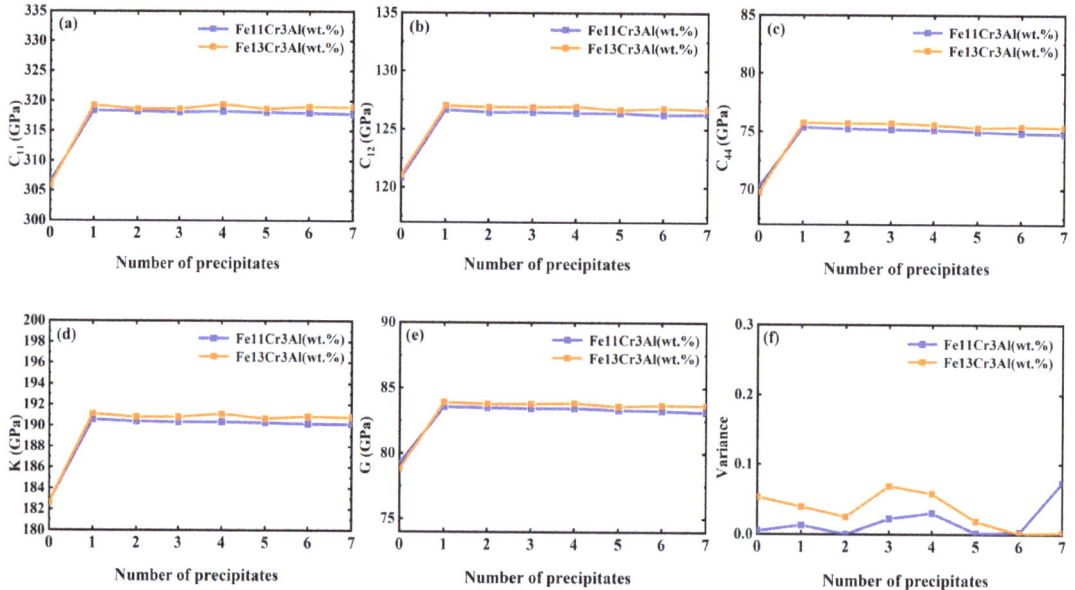

Figure 7. (a–e) The dependence of C_{11}, C_{12}, C_{44}, bulk modulus and shear modulus on number of Cr precipitates. (f) Example of statistical uncertainty for these elastic constants based on 50 simulations at 0 K.

4. Conclusions

In this work, the elastic constants, C_{11}, C_{12}, C_{44}, bulk modulus and shear modulus of FeCrAl alloys were calculated with molecular dynamics methods for compositions with 1~15 wt.% Cr and 1~5 wt.% Al at temperatures from 0 K to 750 K. The effects of Cr and Al concentrations and temperature on elastic constants have been explored. The results indicate that C_{11}, C_{12}, C_{44}, bulk modulus and shear modulus show different dependencies on the concentrations of Cr and Al. In particular, for alloys with Al concentration >3 wt.%, the bulk modulus decreases with the increase of Cr concentration. However, for Al concentration of 1–5 wt.%, the shear modulus decreases with the increase of Cr concentration. Both results are consistent with experimental results. An decrease of elastic constants with increasing temperature was also observed in the present work, which is consistent with experimental results. Investigation on elastic properties of defect-containing alloys have shown that vacancies, voids, interstitials and Cr-rich precipitations have different effects on elastic properties of FeCrAl alloys. Vacancies and voids result in the decrease of elastic properties, while interstitials could result in the increase of bulk modulus but decrease of shear modulus. The formation of Cr-rich precipitates increases the elastic properties, which is independent of the number and volume of precipitates. Therefore, the present results indicate that compared to the concentration effect, the radiation defects show different effects on elastic properties. Furthermore, when the radiation effects are predicted, e.g., radiation hardening, not only the contribution from the interaction between dislocation and radiation defects, but also the radiation defects before the interactions with dislocations, should be considered. All these results provide a new understanding to develop and apply the FeCrAl alloy in AFT design.

Author Contributions: Conceptualization, N.G. and X.W.; methodology, H.D.; software, M.Y.; validation, M.Y. and N.G.; formal analysis, H.D. and M.Y.; investigation, H.D. and M.Y.; resources, H.D.; data curation, Y.D.; writing—original draft preparation, H.D.; writing—review and editing, N.G. and W.S.; visualization, H.D.; supervision, N.G.; project administration, N.G.; funding acquisition, N.G. All authors have read and agreed to the published version of the manuscript.

Funding: This work was funded by the National Natural Science Foundation of China (Grant number. 12075141 and 11675230). WS acknowledges the support provided by the U.S. Department of Energy through the Office of Fusion Energy Sciences (DE-AC05-76RLO-1830).

Institutional Review Board Statement: Not applicable.

Informed Consent Statement: Not applicable.

Data Availability Statement: The data presented in this study are available on request from the corresponding author.

Conflicts of Interest: The authors declare no conflict of interest.

References

1. Badini, C.; Laurella, F. Oxidation of FeCrAl alloy: Influence of temperature and atmosphere on scale growth rate and mechanism. *Surf. Coat. Technol.* **2001**, *135*, 291–298. [CrossRef]
2. Han, X.; Wang, Y.; Peng, S.; Zhang, H. Oxidation behavior of FeCrAl coated Zry-4 under high temperature steam environment. *Corros. Sci.* **2019**, *149*, 45–53. [CrossRef]
3. Briggs, S.A.; Edmondson, P.D.; Littrell, K.C.; Yamamoto, Y.; Howard, R.H.; Daily, C.R.; Terrani, K.A.; Sridharan, K.; Field, K.G. A combined APT and SANS investigation of α′ phase precipitation in neutron-irradiated model FeCrAl alloys. *Acta Mater.* **2017**, *129*, 217–228. [CrossRef]
4. Park, D.J.; Kim, H.G.; Park, J.Y.; Jung, Y.I.; Park, J.H.; Koo, Y.H. A study of the oxidation of FeCrAl alloy in pressurized water and high-temperature steam environment. *Corros. Sci.* **2015**, *94*, 459–465. [CrossRef]
5. Wu, X.; Kozlowski, T.; Hales, J.D. Neutronics and fuel performance evaluation of accident tolerant FeCrAl cladding under normal operation conditions. *Ann. Nucl. Energy* **2015**, *85*, 763–775. [CrossRef]
6. Ukai, S.; Kato, S.; Furukawa, T.; Ohtsuka, S. High-temperature creep deformation in FeCrAl-oxide dispersion strengthened alloy cladding. *Mater. Sci. Eng. A* **2020**, *794*, 139863. [CrossRef]
7. Terrani, K.A.; Zinkle, S.J.; Snead, L.L. Advanced oxidation-resistant iron-based alloys for LWR fuel cladding. *J. Nucl. Mater.* **2014**, *448*, 420–435. [CrossRef]

8. Altıparmak, S.C.; Yardley, V.A.; Shi, Z.; Lin, J. Challenges in additive manufacturing of high-strength aluminium alloys and current developments in hybrid additive manufacturing. *Int. J. Lightweight Mater. Manuf.* **2021**, *4*, 246–261. [CrossRef]
9. Kobayashi, S.; Takasugi, T. Mapping of 475 °C embrittlement in ferritic Fe–Cr–Al alloys. *Scr. Mater.* **2010**, *63*, 1104–1107. [CrossRef]
10. Edmondson, P.D.; Briggs, S.A.; Yamamoto, Y.; Howard, R.H.; Sridharan, K.; Terrani, K.A.; Field, K.G. Irradiation-enhanced α′ precipitation in model FeCrAl alloys. *Scr. Mater.* **2016**, *116*, 112–116. [CrossRef]
11. Chang, K.; Meng, F.; Ge, F.; Zhao, G.; Du, S.; Huang, F. Theory-guided bottom-up design of the FeCrAl alloys as accident tolerant fuel cladding materials. *J. Nucl. Mater.* **2019**, *516*, 63–72. [CrossRef]
12. Gussev, M.N.; Field, K.G.; Yamamoto, Y. Design, properties, and weldability of advanced oxidation-resistant FeCrAl alloys. *Mater. Des.* **2017**, *129*, 227–238. [CrossRef]
13. Thompson, Z.T.; Terrani, K.A.; Yamamoto, Y. *Elastic Modulus Measurement of ORNL ATF FeCrAl Alloys*; ORNL/TM-2015/632; Materials Science: Oak Ridge, TN, USA, 2015; pp. 1–17.
14. Liu, Z.; Li, Y.; Shi, D.; Guo, Y.; Li, M.; Zhou, X.; Huang, Q.; Du, S. The development of cladding materials for the accident tolerant fuel system from the Materials Genome Initiative. *Scr. Mater.* **2017**, *141*, 99–106. [CrossRef]
15. Busby, J.T.; Hash, M.C.; Was, G.S. The relationship between hardness and yield stress in irradiated austenitic and ferritic steels. *J. Nucl. Mater.* **2005**, *336*, 267–278. [CrossRef]
16. Speich, G.; Schwoeble, A.; Leslie, W.C. Elastic constants of binary iron-base alloys. *Metall. Trans.* **1972**, *3*, 2031–2037. [CrossRef]
17. Hellier, A.; Palmer, S.; Whitehead, D. An integrated circuit pulse echo overlap facility for measurement of velocity of sound (applied to study of magnetic phase change). *J. Phys. E Sci. Instrum.* **1975**, *8*, 352. [CrossRef]
18. Harmouche, M.; Wolfenden, A. Modulus measurements in ordered Co-Al, Fe-Al, and Ni-Al alloys. *J. Test. Eval.* **1985**, *13*, 424–428.
19. Field, K.G.; Briggs, S.A.; Hu, X.; Yamamoto, Y.; Howard, R.H.; Sridharan, K. Heterogeneous dislocation loop formation near grain boundaries in a neutron-irradiated commercial FeCrAl alloy. *J. Nucl. Mater.* **2017**, *483*, 54–61. [CrossRef]
20. Field, K.G.; Briggs, S.A.; Sridharan, K.; Howard, R.H.; Yamamoto, Y. Mechanical Properties of Neutron-Irradaited Model and Commercial FeCrAl Alloys. *J. Nucl. Mater.* **2017**, *489*, 118–128. [CrossRef]
21. Garner, F.; Toloczko, M.; Sencer, B. Comparison of swelling and irradiation creep behavior of fcc-austenitic and bcc-ferritic/martensitic alloys at high neutron exposure. *J. Nucl. Mater.* **2000**, *276*, 123–142. [CrossRef]
22. Liao, X.; Gong, H.; Chen, Y.; Liu, G.; Liu, T.; Shu, R.; Liu, Z.; Hu, W.; Gao, F.; Jiang, C.; et al. Interatomic potentials and defect properties of Fe–Cr–Al alloys. *J. Nucl. Mater.* **2020**, *541*, 152421. [CrossRef]
23. Miller, A.; Saunders, G.; Yogurtcu, Y. Pressure dependences of the elastic constants of PbTe, SnTe and $Ge_{0.08}Sn_{0.92}$Te. *J. Phys. C Solid State Phys.* **1981**, *14*, 1569. [CrossRef]
24. Xu, S.; Xie, D.; Liu, G.; Ming, K.; Wang, J. Quantifying the resistance to dislocation glide in single phase FeCrAl alloy. *Int. J. Plast.* **2020**, *132*, 102770. [CrossRef]
25. Voight, W. *Lehrbuch der Kristallphysik*; Teubner: Leipzig, Germany, 1928.
26. Yeom, H.; Maier, B.; Johnson, G.; Dabney, T.; Walters, J.; Sridharan, K. Development of cold spray process for oxidation-resistant FeCrAl and Mo diffusion barrier coatings on optimized ZIRLO^{TM}. *J. Nucl. Mater.* **2018**, *507*, 306–315. [CrossRef]
27. Yamamoto, Y.; Sun, Z.; Pint, B.A.; Terrani, K.A. *Optimized Gen-II FeCrAl Cladding Production in Large Quantity for Campaign Testing*; Oak Ridge National Laboratory: Oak Ridge, TN, USA, 2016.
28. Wang, R.; Zhang, X.; Wang, H.; Ni, J. Phase diagrams and elastic properties of the Fe-Cr-Al alloys: A first-principles based study. *Calphad* **2019**, *64*, 55–65. [CrossRef]
29. Leibfried, G.; Ludwig, W. *Theory of Anharmonic Effects in Crystals, Solid State Physics*; Elsevier: Amsterdam, The Netherlands, 1961; pp. 275–444.
30. Li, D.Z.; Zhang, X.D.; Li, J.; Zhao, L.J.; Wang, F.; Chen, X.Q. Insight into the elastic anisotropy and thermodynamics properties of Tantalum borides. *Vacuum* **2019**, *169*, 108883. [CrossRef]

Article

Atomic Simulations of the Interaction between a Dislocation Loop and Vacancy-Type Defects in Tungsten

Linyu Li [1], Hao Wang [2], Ke Xu [1], Bingchen Li [1], Shuo Jin [1], Xiao-Chun Li [3], Xiaolin Shu [1], Linyun Liang [1,*] and Guang-Hong Lu [1,*]

[1] School of Physics, Beihang University, Beijing 100191, China; linyu@buaa.edu.cn (L.L.); xuuke@buaa.edu.cn (K.X.); by1919006@buaa.edu.cn (B.L.); jinshuo@buaa.edu.cn (S.J.); shuxlin@buaa.edu.cn (X.S.)

[2] Center for Fusion Science, Southwestern Institute of Physics, Chengdu 610041, China; hwang@swip.ac.cn

[3] Institute of Plasma Physics, Chinese Academy of Sciences, Hefei 230031, China; xcli@ipp.ac.cn

* Correspondence: lyliang@buaa.edu.cn (L.L.); lgh@buaa.edu.cn (G.-H.L.)

Abstract: Tungsten (W) is considered to be the most promising plasma-facing material in fusion reactors. During their service, severe irradiation conditions create plenty of point defects in W, which can significantly degrade their performance. In this work, we first employ the molecular static simulations to investigate the interaction between a 1/2[111] dislocation loop and a vacancy-type defect including a vacancy, di-vacancy, and vacancy cluster in W. The distributions of the binding energies of a 1/2[111] interstitial and vacancy dislocation loop to a vacancy along different directions at 0 K are obtained, which are validated by using the elasticity theory. The calculated distributions of the binding energies of a 1/2[111] interstitial dislocation loop to a di-vacancy and a vacancy cluster, showing a similar behavior to the case of a vacancy. Furthermore, we use the molecular dynamics simulation to study the effect of a vacancy cluster on the mobility of the 1/2[111] interstitial dislocation loop. The interaction is closely related to the temperature and their relative positions. A vacancy cluster can attract the 1/2[111] interstitial dislocation loop and pin it at low temperatures. At high temperatures, the 1/2[111] interstitial dislocation loop can move randomly. These results will help us to understand the essence of the interaction behaviors between the dislocation loop and a vacancy-type defect and provide necessary parameters for mesoscopic scale simulations.

Keywords: atomic simulations; dislocation loop; vacancy defect; tungsten

1. Introduction

Tungsten (W), owing to its high melting temperature, good thermal conductivity, and low sputtering yield, is believed to be the most promising candidate for plasma-facing materials (PFMs) in fusion reactors [1–4]. During the operation of the reactors, high-energy neutrons escaped from the plasma will bombard on PFMs, creating plenty of self-interstitial atoms (SIAs) and vacancies in them [5]. These point defects further aggregate into small vacancy clusters including voids and dislocation loops [6–10]. Experimental observations showed voids and dislocation loops are major defect clusters in pure W at the low dpa level (less than 1.54) in the temperature range from 300 to 900 °C [11]. The dislocation loop can be either the vacancy dislocation loop or the interstitial dislocation loop with different properties [10–13]. The existence of various irradiated defects makes the system difficult to be understood. The interactions between them further complex the system, which is believed to play an important role in the microstructural evolution of PFMs. For examples, the dislocation loop can grow or shrink by absorbing SIAs and vacancies, respectively [14,15]. Two dislocation loops attract or repel each other when they are in different relative positions [16]. The dislocation loops and voids act as an obstacle to the dislocation glide [17–23]. For the interaction between the dislocation loop and the dislocation, it closely depends on the character and nature of the loop, and their relative positions and temperature [16,18–20]. For the interaction between voids and dislocations, the presence of an edge dislocation in the vicinity of the void can generate a stress field that impacts on the motion of the dislocation [23]. Therefore, understanding the interaction between these defects is very important to explore their microscopic mechanism and thus correlate them to the macroscopic mechanical properties [10,16,20,21,23].

Several previous studies focused on the interaction behaviors between vacancy-type defects and dislocation loops in iron [23–26]. Dislocation loops act as biased sinks that attract SIAs and vacancies. For the interaction between a vacancy and SIA clusters (9-127SIAs), the vacancy can be annihilated only when it is placed along the edge of the dislocation loop and parallel to the Burgers vector. When the vacancy is in a site next to the center of the cluster, it does not annihilate with SIAs but affects the motion of the cluster, reducing or even preventing its migration [24]. Furthermore, their interactions are temperature and cluster size dependent. Thus, the vacancy has an influence on the movement and evolution of the dislocation loop. When the vacancies aggregate into vacancy clusters, they can further interact with the dislocation loop in a large distance. The vacancy clusters can attract the dislocation loop due to the elastic interaction between them [27]. Molecular dynamics (MD) simulations showed that when the vacancy cluster is placed within the interaction distance to a 1/2[111] dislocation loop in W, it directly diffuses towards the vacancy cluster. The diffusion speed of the dislocation loop is related to the shape, size, and position of the vacancy clusters. The vacancy clusters can be annihilated eventually if the dislocation loop is large enough [27]. However, although several studies of the interaction between vacancy-defects and dislocation loops have been done for iron, there is still a lack of relevant investigations on W, in which the stable structure of the interstitial crowdion and the diffusion mechanism of the dislocation loop are different from that of ion [28,29]. Therefore, the objective of this work is to systematically investigate the interaction between a dislocation loop and a vacancy-type defect in W.

In this work, we study the interaction behaviors of the dislocation loop and a vacancy, di-vacancy, and vacancy clusters by using the molecular statics (MS) and MD simulations. We first calculate the binding energies of two types of 1/2[111] dislocation loops to a monovacancy. To validate the simulation results, the widely-used elasticity theory (ET) is performed to calculate their binding energies [24–26]. Then we calculate the binding energies of the 1/2[111] interstitial dislocation loop to a divacancy and a vacancy cluster using MS simulations. The effect of the vacancy cluster on the mobility of the 1/2[111] dislocation loop at different temperatures is also investigated using MD simulations. We hope our results can provide useful datasets for large-scale simulations such as kinetic Monte Carlo, cluster dynamics, and dislocation dynamics and help to study the long-term and large-scale microstructure evolution in W under irradiation.

2. Simulation Details

We employ the Large-scale Atomic/Molecular Massively Parallel Simulator (LAMMPS) software (Sandia National Laboratories, Albuquerque, NM, USA) to study the interaction between a dislocation loop and vacancy-type defects including a vacancy, di-vacancy, and vacancy cluster. A previous study indicated that nearly 60% of the dislocation loop is the 1/2[111] dislocation loop in W at 0.4–30 dpa and temperatures between 300 and 750 °C [21]. The size of the dislocation loop is usually less than 20 nm with most of them less than 6 nm [22]. Thus, we choose the 1/2[111] dislocation loop with its radius less than 6 nm. The embedded atom method (EAM) empirical interatomic potential developed by Marinica et al., denoted as "EAM2", is employed in our simulations, which has been used to investigate the properties of dislocation loop [30,31].

The direction of X, Y, and Z is set as [$\bar{1}\bar{1}2$], [$1\bar{1}0$], and [111], respectively. The simulation box created by LAMMPS contains about 0.59 million atoms. Periodic boundary conditions are imposed on all boundaries of the simulation box. The 1/2[111] dislocation loop is placed at the center of the box and the vacancy-type defects are created at designed positions. The sketch of the simulation box and a schematic picture of the configuration of a 1/2[111] dislocation loop and a vacancy-type defect is shown in Figure 1.

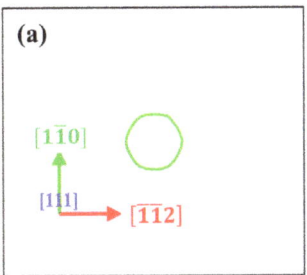
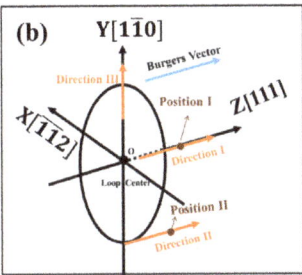

Figure 1. (**a**) The sketch of the simulation box. The green ring represents the 1/2[111] dislocation loop. (**b**) The schematic picture of the configuration of a 1/2[111] dislocation loop and a vacancy-type defect. The vacancy-type defect is placed at the position that is along three different directions. Direction I is through the center of the dislocation loop and parallel to the Burgers vector. Position I is along the Direction I and has a distance of 15 Å to the habit plane (HP). Direction II is along the edge of the dislocation loop and parallel to the Burgers vector. Position II is along the Direction II and has the distance of 15 Å to the HP. Direction III is through the center of the dislocation loop on the HP.

Binding energies can be used to evaluate the static interaction between a dislocation loop and a vacancy-type defect. We first construct a 1/2[111] dislocation loop at the center of the simulation box. Then we insert a vacancy-type defect at a certain position (Position I and II) as shown in Figure 1b and relax the system to reach the equilibrium state by using the conjugate gradient method. The binding energy of the loop to a vacancy-type defect is given by:

$$E_b = E_1 - E_2 \qquad (1)$$

where E_1 stands for the minimum energy of the system with a vacancy-type defect that is far away from the dislocation loop to ensure that there is no interaction between them, and E_2 stands for the minimum energy of the system with a vacancy-type defect that is in a specific position. A positive E_b represents the attraction of the dislocation loop to a vacancy-type defect, while the negative value denotes the repulsion between them.

The elasticity theory is first used to verify our MS simulation results. A vacancy can be regarded as a sphere whose elastic constants are zero. On the one hand, the elastic constant of a vacancy is different from that of system, which will interfere with the existing elastic field and produce the interaction energy E_{int}^1. On the other hand, the stress produced by the vacancy will interact with the original elastic field and produce the interaction energy E_{int}^2. According to the linear elasticity theory, the total interaction energy is equal to the sum of the two energies E_{int}^1 and E_{int}^2 [32]. Then the interaction energy between a dislocation loop and a vacancy can be obtained. The HP of the dislocation loop is (r, θ) with a radius R. The center of the loop is at the origin and the Burgers vector is parallel to the Z axis. The interaction energy between a vacancy and a loop can be calculated by the following formula [32]:

$$E_{int}^1(\xi, \rho) = -K^2(1-\sigma)\Omega\left\{\frac{1}{3\kappa}\frac{(1+\sigma)^2}{1-2\sigma}I_0^{12} + \frac{15}{2\mu}\frac{1}{7-5\sigma}\left[\frac{(1-2\sigma)^2}{3}\left(I_0^1\right)^2 + \xi^2 I_0^{2} + \xi^2 I_1^{2} + \rho^{-2}\Phi^2 + \xi\rho^{-1}I_0^2\Phi - (1-2\sigma)\rho^{-1}I_0^1\Phi\right]\right\}, \qquad (2)$$

$$E_{int}^2(\xi, \rho) = -\frac{2}{3}K(1+\sigma)\Delta\Omega I_0^1 \qquad (3)$$

where:

$$\xi = \frac{z}{R} \qquad (4)$$

$$\rho = \frac{r}{R} \qquad (5)$$

$$K = \frac{b}{R} \frac{\mu}{2(1-\sigma)} \quad (6)$$

$$I_m^n(\xi, \rho) = \int_0^\infty t^n J_m(t\rho) J_1(t) e^{-\xi t} dt \, (n, m = 0, 1, 2) \quad (7)$$

$$\Phi(\xi, \rho) = (1 - 2\sigma) I_0^1 - \xi I_1^1 \quad (8)$$

where (z, r) is the position of the vacancy, R is the dislocation loop radius, k and μ represents the elastic constant tensor C_{12} and C_{44} of W, respectively. k_1 and μ_1 represent the elastic constant of the vacancy, which equals 0. σ is the Poisson's ratio. Ω is the volume of the vacancy, and $I_m^n(\xi, \rho)$ is a complete elliptic integral. The used elastic constant tensors of W are $C_{11} = 523$ GPa, $C_{12} = 203$ GPa, and $C_{44} = 160$ Gpa.

3. Results and Discussion

We first use the MS to calculate the binding energies of the 1/2[111] interstitial dislocation loop (IDL) and 1/2[111] vacancy dislocation loop (VDL) to a vacancy by varying the type and size of the loop at 0 K. The binding energies of the IDL to a vacancy are compared with that calculated by ET. We then simulate the binding energies of the IDL to a di-vacancy and a vacancy cluster. Finally, the effects of the temperature and position of the vacancy cluster on the mobility of the IDL are obtained by the MD simulations.

3.1. Interaction between an 1/2[111] IDL to a Vacancy

Two types of the 1/2[111] dislocation loop, IDL and VDL, can be experimentally observed [21]. To understand the interaction mechanism of the IDL and VDL to a vacancy, we calculate their binding energies by varying the relative position of the vacancy. Figure 2 depicts the binding energies of the IDL to a vacancy on the HP of the loop along the Direction I and Direction II.

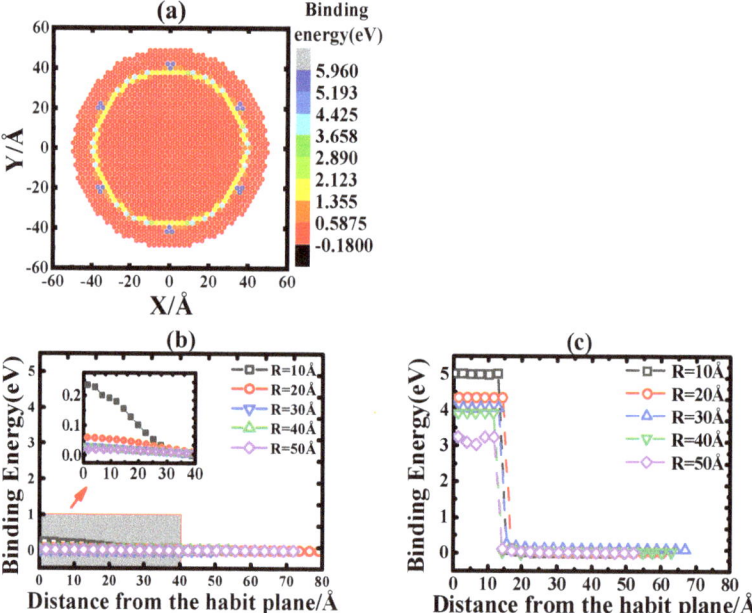

Figure 2. The distribution of binding energies of an IDL to a vacancy when the vacancy is placed on the HP (**a**), along the Direction I (**b**), and along the Direction II (**c**), respectively. The inset figure shows the binding energies at a small energy scale.

The distribution of the binding energies of IDL (R = 40.0 Å) to a vacancy placed on the HP (XY plane) is shown in Figure 2a. We choose the center of the loop as the origin of the coordinate. The binding energy is calculated by creating a vacancy at every lattice position on the HP. The color bar indicates the values of the binding energies of the IDL to the vacancy. It can be seen that the binding energy is relatively small when the vacancy is placed far away from the dislocation loop. The binding energy becomes large when the vacancy is placed close to the edge of the IDL, where the vacancy can be absorbed. We define the regime that the IDL can absorb the vacancy as their absorption area. This absorption area has a ring-like shape and its width is defined as the absorption distance. The calculated absorption distance is ~8.0 Å. The binding energies approach zero when the vacancy is not in the absorption area.

Due to the symmetrical distribution of the binding energies on the HP as shown in Figure 2a, we calculate the binding energies only along one particular direction to save computational resources. Figure 2b shows the calculated distribution of binding energies of the IDL to a vacancy as a function of their distance along the Direction I. With the increase of the distance, the binding energies gradually decrease for the same sized IDL. The binding energies are almost zero (<0.025 eV) when the distance is larger than 30.0 Å for all sized IDL. As the radius of the IDL is increased from 10.0 Å to 50.0 Å, the highest binding energy decreases from 0.25 eV to 0.01 eV. Figure 2c shows the distribution of the binding energies between the IDL and vacancy as a function of their distance along the Direction II. The binding energies remain constants within a certain distance for the same sized IDL. This indicates the vacancy is absorbed by IDL. Beyond this distance, the binding energies are nearly zero (<0.15 eV). This distance is defined as the absorption distance along the Direction II. With the radius of the IDL increasing from 10.0 Å to 50.0 Å, the largest binding energy decreases from 5.1 eV to 1.4 eV and the absorption distance is 14.0–17.0 Å. There is no clear evidence that the IDL size has an influence on the absorption distance based on our simulation results.

Comparing the binding energies calculated along the Direction I and Direction II, we can find that the binding energies along the Direction II are always larger than that along the Direction I for the same sized IDL with the same relative distance. This implies that the attraction of the IDL to the vacancy is stronger along the Direction II than that of the Direction I. The vacancy can be annihilated by IDL only when it is placed close to the edge of the loop.

We find that a stable VDL cannot be formed in the system if the radius of the VDL is smaller than 40.0 Å in our simulations. It will evolve into a void. Therefore, we construct a series of VDLs with a radius of 40.0 Å to 60.0 Å. Figure 3 shows the distribution of the binding energies of VDL to a vacancy on the HP of the loop along the Direction I and Direction II, respectively.

The distribution of the calculated binding energies of the VDL (40.0 Å) to a vacancy placed on the HP is shown in Figure 3a. It shows the absorption area of the VDL to a vacancy also has a ring-like shape, which is very similar to the case of the IDL to a vacancy. The largest binding energy is around 2.2 eV, which is smaller than that of the IDL to a vacancy. The binding energies approach to zero when the vacancy is placed outside the absorption area.

Figure 3b depicts the distribution of the binding energies as a function of their distance along the Direction I. The binding energies are negative for all sized VDL, implying the repulsive interaction between the VDL and the vacancy. The largest binding energy increases from −0.032 eV to −0.02 eV with the radius of the VDL increasing from 40.0 Å to 60.0 Å. As the distance increases, the value of the binding energy increases for the same sized VDLs. If the distance is larger than 30.0 Å, the value of the binding energy is less than 0.015 eV. The distribution of the binding energies as a function of their distance along the Direction II is shown in Figure 3c. The largest binding energy decreases from 2.4 eV to 0.9 eV with the increase of the radius of the loop from 40.0 Å to 60.0 Å. Within the absorption distance (~15.0 Å), the binding energy of a VDL to a vacancy decreases gradually that is different from the case of the IDL and a vacancy. Beyond this absorption distance, the binding energies are nearly zero.

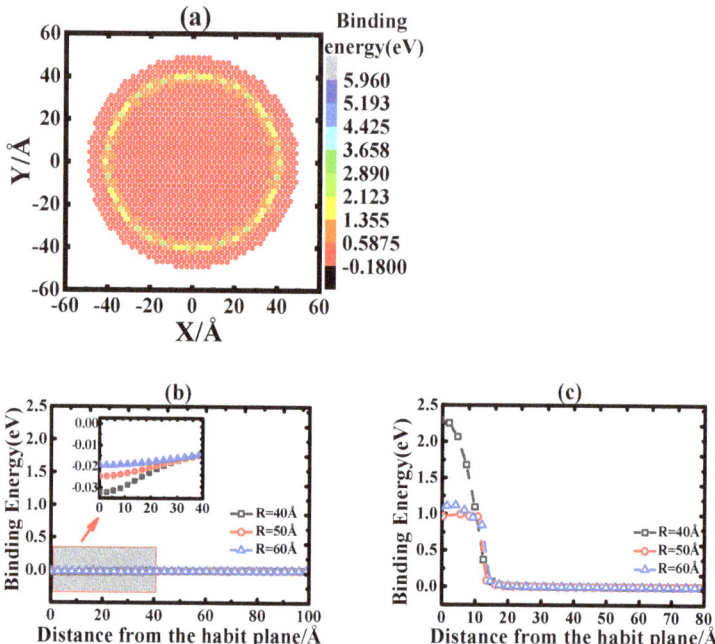

Figure 3. The distribution of binding energies of a VDL to a vacancy when the vacancy is placed on (**a**) HP, along (**b**) the Direction I and (**c**) Direction II, respectively. The inset figure shows the binding energies at a small energy scale.

Based on the above simulation results, the distribution of the binding energies of a VDL to a vacancy are similar to that of an IDL to a vacancy. However, we find that the binding energies of the IDL to a vacancy are positive while the binding energies of the VDL to a vacancy are negative. Thus, the IDL can attract the vacancy while the VDL can slightly repulse it along the Direction I. The reason can be explained by the distribution of the stress for different dislocation loops. As shown in Figure 4a, there is a compressive stress inside the IDL but a tensile stress outside it. While the signs of the stress are opposite for the VDL as shown in Figure 4b. A vacancy is very easily combined with the IDL, which is the main reason to have positive binding energies inside the IDL. The different distribution of the stress contributes to the different interaction behaviors of the IDL and VDL to a vacancy.

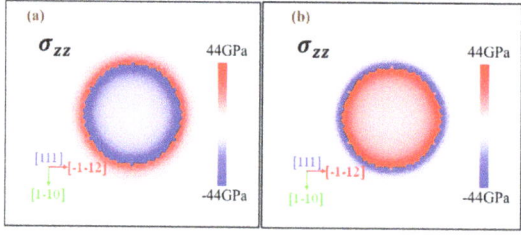

Figure 4. The stress distributions of the (**a**) interstitial dislocation loop and (**b**) vacancy dislocation loop with a radius of 40.0 Å.

3.2. Comparison of the Binding Energies of IDL-Vacancy by Using ET and MS

To better understand the interaction behaviors between a vacancy and a 1/2[111] IDL, we compare their binding energies with different relative positions calculated by using both MS and ET as shown in Figure 5. The radius of the IDL is chosen as 40.0 Å as an example. The vacancy is placed at several selected positions along the Direction II and Direction III.

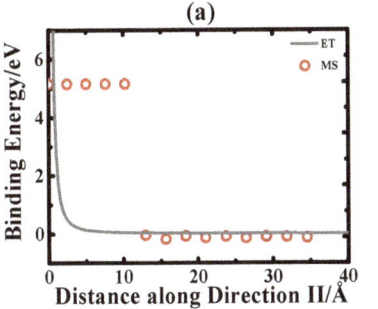
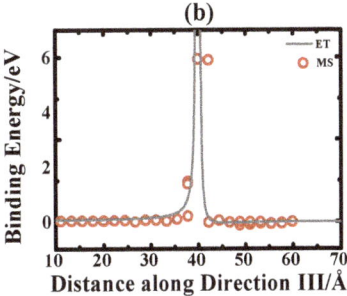

Figure 5. Comparison of the binding energies of the IDL (40.0 Å) to a vacancy calculated by MS and ET along (**a**) the Direction II and (**b**) the Direction III, respectively.

Figure 5a shows the comparison of the binding energies between an IDL and a vacancy calculated by ET and MS as a function of their distance along the Direction II. When the distance is less than 14.0 Å, the binding energies calculated by ET decrease dramatically from an infinity value to zero with increasing the distance, while the binding energies calculated by MS are nearly a constant. The difference can be attributed to the ET cannot accurately describe the stress field when two defects are close to each other, and also the vacancy is completely absorbed by IDL in MS simulations that is not the case in ET. Beyond this distance, the binding energies given by ET and MS are in good agreements. Figure 5b shows the comparison of the binding energies calculated by the ET and MS as a function of their distance along the Direction III. Based on the difference of the binding energies between the IDL and a vacancy, we can categorize the distribution of the binding energies into two different regimes, inside the absorption distance that is greater than 35.0 Å and less than 42.0 Å and beyond the absorption distance. Within the absorption distance, the binding energies predicted by ET tend to be infinity, while the binding energy given by MS is 5.93 eV. Beyond the absorption distance, the binding energies calculated by ET agree well with that given by MS. Furthermore, the stress inside the loop is larger than that outside the loop as shown in Figure 4a. Thus, it can be seen from Figure 5b that the binding energies given by ET are slightly larger for the vacancy placed inside the loop than that outside the loop along the Direction III. In conclusion, we basically validate our MS simulation results although the ET can not fully describe the interaction between an IDL and a vacancy.

3.3. Static Interaction of an IDL to a Di-Vacancy and a Vacancy Cluster

To investigate the interaction between the IDL to a di-vacancy and a vacancy cluster, we calculate their binding energies along the Direction II and Direction III, respectively. The configuration of the divacancy considered is 1NN, which was reported stable in W [6]. Figure 6 shows the distribution of the calculated binding energies between the different sized IDLs and a di-vacancy as a function of their distance along the Direction I and II, respectively. The variations of the binding energies with the distance and loop radius are very similar to that of an IDL and a vacancy as shown in Figure 2. The largest binding energy of the IDL to a divacancy is generally larger than that of the IDL to a vacancy. Along the Direction I as shown in Figure 6a the largest binding energy decreases from 0.45 eV to 0.03 eV with increasing the radius of the IDL from 10.0 Å to 50.0 Å. The binding energy reaches zero when the distance is larger than 40.0 Å for all sized IDLs. While along the

Direction II as shown in Figure 6b, the largest binding energy decreases from 5.8 eV to 1.2 eV with increasing the radius of the loop from 10.0 Å to 50.0 Å. The calculated absorption distance ranges from 15.0 Å to 27.0 Å.

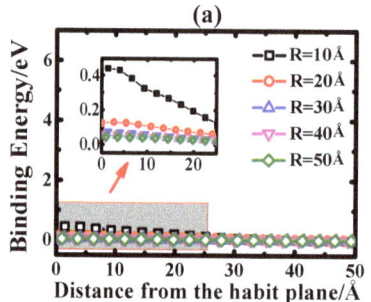 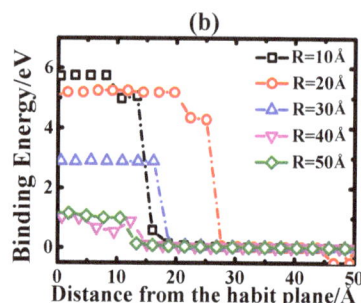

Figure 6. The distribution of binding energies of the different sized IDLs to a divacancy as a function of their distance when the divacancy is placed along (**a**) the Direction I and (**b**) Direction II, respectively. The inset figure shows the binding energies at a small energy scale.

Figure 7 depicts the calculated binding energies between the different sized IDLs and a vacancy cluster as a function of their distance along the Direction I and II. Comparing to the binding energies of the IDL to a vacancy and a di-vacancy as shown in Figures 5 and 6, the distribution of the binding energies of IDL to a vacancy cluster as a function of their distance shows a similar trend but larger values. Figure 7a shows the binding energy decreases with the increase of their distance and the loop radius along the Direction I. The largest binding energy decreases from 3.4 eV to 0.05 eV with increasing the radius of the loop from 10.0 Å to 50.0 Å. The binding energy is almost zero when the distance is longer than 40.0 Å for all sized IDLs. While along the Direction II as shown in Figure 7b, the largest binding energy decreases from 12.0 eV to 6.2 eV as the IDL radius increases from 10.0 Å to 50.0 Å. The absorption distance ranges from 22.0 Å to 32.0 Å. Therefore, as the number of vacancies increases, the binding energies increase gradually for the same sized IDL and the same distance. Besides, the distribution of the binding energies of the IDL to various vacancy-type defects as a function of the distance along the Direction I and II is similar. Based on the same trend of the distribution of the binding energies between the IDL and vacancy-type defects, we can conclude that they have similar interaction behaviors.

3.4. Dynamic Interaction between an IDL and a Vacancy Cluster

A previous study showed that the IDL exhibits a fast one-dimensional motion along the <111> direction in W [16]. The existence of a vacancy inside the IDL can inhibit its motion, which also closely depends on the IDL size and temperature [24]. Besides, the number of the vacancies can also have a large influence on the movement of the IDL or even absorb it [27]. Therefore, the vacancy cluster plays an important role in the evolution of the IDL and needs to be investigated systemically.

We calculate the movement distance of the IDL (10.0 Å) along the Burgers vector with a vacancy cluster (containing eight vacancies) by varying its position and temperature. The movement distances of the IDL along the Burgers vector as a function of the simulation time at different temperatures are shown in Figure 8. The inset pictures show the movement distance of the IDL at the initial stage (0–150 ps). The interaction between the IDL and the vacancy cluster largely depends on the temperature.

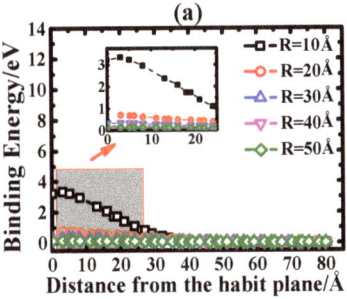
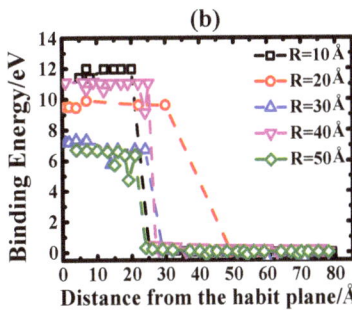

Figure 7. The distribution of binding energies of the different sized IDLs to a vacancy cluster (containing eight vacancies) as a function of their distance when the vacancy clusters is placed along the (**a**) Direction I and (**b**) Direction II, respectively. The inset figure shows the binding energies at a small energy scale.

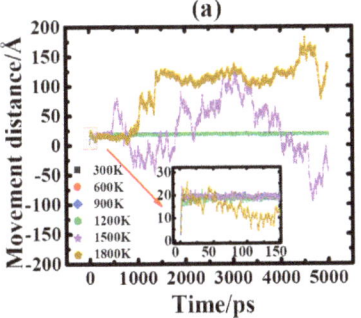
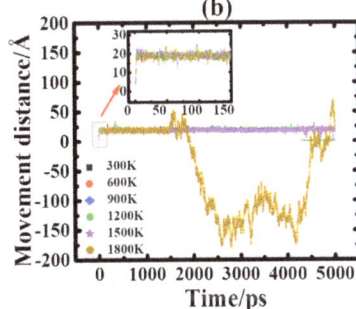

Figure 8. The movement distance of the IDL along the Burgers vector when a vacancy cluster is placed at (**a**) Position I and (**b**) Position II, respectively.

Figure 8a shows the movement distance of the IDL with the vacancy cluster placed at the position I as a function of the simulation time at different temperatures. We can see that the IDL will diffuse towards the vacancy cluster at the initial stage as shown in the insert figure. Then the IDL is captured by the vacancy cluster for all temperatures. At a temperature below 1200 K, the simulation results show the IDL is pinned by the vacancy cluster and immobile. At a temperature above 1500 K, the IDL can move after 400 ps and 800 ps, respectively. Based on the analysis of the evolution of the microstructure, the vacancy cluster first attracts the IDL to drive it to the position of the vacancy cluster, then the vacancies inside the cluster will be absorbed by IDL. However, not all vacancies in the vacancy cluster will be absorbed. The left vacancies will inhibit the movement of the IDL. At a temperature above 1200 K, the vacancy captured inside the IDL can migrate towards the edge of the IDL, and then the vacancy can be annihilated. Finally, the loop mobility is recovered. Figure 8b shows the calculated movement distance of the IDL with a vacancy cluster placed at Position II at several selected times. It is shown that the interaction behavior is very similar to that with a vacancy cluster placed at Position I. While the IDL requires a higher temperature of 1800K than that at Position I to unpin the vacancy cluster.

In conclusion, the movement distance of the IDL depends on the position of the vacancy cluster and temperature. At low temperature, the IDL is pinned by the vacancy cluster, while it can move at high temperature because it can promote the recombination of the IDL and the vacancy. Besides, the attracting capability of the IDL by the vacancy cluster placed at Position II is stronger than it placed at Position I.

4. Conclusions

In this work, we employ an atomic simulation to investigate the interaction behaviors between a dislocation loop and vacancy-type defects including a vacancy, di-vacancy, and vacancy cluster. The distribution of the binding energies of an 1/2[111] interstitial dislocation loop (IDL) to a vacancy is different from that of a 1/2[111] vacancy dislocation loop (VDL) to a vacancy. When the vacancy is adjacent to the center of the loop, the VDL repulses the vacancy, but the IDL attracts it due to their different stress distributions. The binding energies calculated by elasticity theory (ET) and molecular statics simulations show that they are consistent when they are far away from each other, but it has a large derivation when they are close due to the ET has a difficulty in accurately predicting the stress near the core of the loop. The interaction behaviors between the IDL and a di-vacancy are very similar to that of the IDL and a vacancy. Furthermore, a vacancy cluster (containing 8 vacancies) can hinder the motion of the IDL, which depends on the temperature and their relative position. No matter where the vacancy cluster is, the IDL absorbs a part of the vacancies in the cluster and the unabsorbed vacancies will inhibit the motion of the dislocation loop at low temperatures. While the mobility of IDL is recovered at high temperature by absorbing all vacancies in nanoseconds. Therefore, the temperature and position dependent interaction of the IDL and a vacancy cluster should be taken into account in modeling the microstructure evolution during irradiation. These obtained binding energies and absorption distances provide input parameters for the kinetic Monte Carlo, cluster dynamics, and dislocation dynamics simulations.

Author Contributions: Conceptualization, G.-H.L., L.L. (Linyun Liang), H.W., K.X. and L.L. (Linyu Li); methodology, G.-H.L., L.L. (Linyun Liang), X.-C.L., H.W., K.X. and L.L.(Linyu Li); software, L.L. (Linyu Li); validation, L.L. (Linyu Li).; formal analysis, G.-H.L., L.L.(Linyun Liang), X.-C.L., H.W., K.X., B.L., S.J., X.S. and L.L. (Linyu Li); software, L.L. (Linyu Li); investigation, L.L. (Linyu Li); resources, L.L. (Linyu Li); data curation, L.L. (Linyu Li); writing—original draft preparation, L.L. (Linyu Li); writing—review and editing, G.-H.L., L.L.(Linyun Liang), X.-C.L., H.W., K.X., B.L., S.J., X.S. and L.L.(Linyu Li); visualization, L.L. (Linyu Li). All authors have read and agreed to the published version of the manuscript.

Funding: This research was funded by the National Natural Science Foundation of China, Grant Numbers. 51871007, 12075021, and 12075023, and the National MCF Energy R&D Program of China, Grant Number No. 2018YFE0308103. The APC was funded by Grant Number No. 2018YFE0308103.

Data Availability Statement: The data presented in this study are available on request from the corresponding author. The data are not publicly available due to technical limitations.

Acknowledgments: The authors acknowledge Xinyue Fan for providing inspiration and dealing with technical problems.

Conflicts of Interest: The authors declare no conflict of interest.

References

1. Bolt, H.; Brendel, A.; Levchuk, D.; Greuner, H. Materials for plasma facing components of fusion reactors. *Energy Mater.* **2006**, *1*, 121–126. [CrossRef]
2. Federici, G.; Skinner, C.; Brooks, J.; Coad, J.; Grisolia, C.; Haasz, A.; Hassanein, A.; Philipps, V.; Pitcher, C.; Roth, J.; et al. Plasma-material interactions in current tokamaks and their implications for next step fusion reactors. *Nucl. Fusion* **2001**, *41*, 1967–2137. [CrossRef]
3. Zinkle, S.; Was, G. Materials challenges in nuclear energy. *Acta Mater.* **2013**, *61*, 735–758. [CrossRef]
4. Ackland, G. Controlling Radiation Damage. *Science* **2010**, *327*, 1587–1588. [CrossRef]
5. Rubia, T.; Soneda, N.; Caturla, M.; Alonso, E. Defect production and annealing kinetics in elemental metals and semiconductors. *J. Nucl. Mater.* **1997**, *251*, 13–33. [CrossRef]
6. Heinola, K.; Djurabekova, F.; Ahlgren, T. On the stability and mobility of di-vacancies in tungsten. *Nucl. Fusion* **2017**, *58*, 026004. [CrossRef]
7. Osetsky, Y.; Bacon, D.; Serra, A.; Singh, B.; Golubov, S. Stability and mobility of defect clusters and dislocation loops in metals. *J. Nucl. Mater.* **1999**, *276*, 65–77. [CrossRef]
8. Jan, F.; Robin, S.; Mason, D.; Nguyen-Manh, D. Nano-sized prismatic vacancy dislocation loops and vacancy clusters in tungsten. *Nucl. Mater. Energy* **2018**, *16*, 60–65.

9. Fikar, J.; Schublin, R. Stability of small vacancy clusters in tungsten by molecular dynamics. *Nucl. Instrum. Methods Phys. Res. B* **2020**, *464*, 56–59. [CrossRef]
10. Ma, P.; Mason, D.; Dudarev, S. Multiscale analysis of dislocation loops and voids in tungsten. *Phys. Rev. Mater.* **2020**, *4*, 103609. [CrossRef]
11. Hasegawa, A.; Fukuda, M.; Nogami, S.; Yabuuchi, K. Neutron irradiation effects on tungsten materials. *Fusion Eng. Des.* **2014**, *89*, 1568–1572. [CrossRef]
12. Yi, X.; Jenkins, M.; Hattar, K.; Edmonodson, P.; Roberts, S. Characterisation of radiation damage in W and W-based alloys from 2 MeV self-ion near-bulk implantations. *Acta Mater.* **2015**, *92*, 163–177. [CrossRef]
13. Yi, X.; Jenkins, M.; Briceno, M.; Roberts, S.; Zhou, Z. In situ study of self-ion irradiation damage in W and W–5Re at 500 °C. *Philos. Mag.* **2013**, *93*, 1715–1738. [CrossRef]
14. Dubinko, V.; Abyzov, A.; Turkin, A. Numerical evaluation of the dislocation loop bias. *J. Nucl. Mater.* **2005**, *336*, 11–21. [CrossRef]
15. Wang, H.; Xu, K.; Wang, D.; Gao, N.; Li, Y.; Jin, S.; Shu, X.; Liang, L.; Lu, G. Anisotropic interaction between self-interstitial atoms and 1/2<111> pdislocation loops in tungsten. *Sci. China Phys.* **2021**, *64*, 257012. [CrossRef]
16. Li, Y.; Boleininger, M.; Robertson, C.; Dupuy, L.; Dudarev, S. Diffusion and interaction of prismatic dislocation loops simulated by stochastic discrete dislocation dynamics. *Phys. Rev. Mater.* **2019**, *3*, 073805. [CrossRef]
17. Shi, S.; Zhu, W.; Huang, H.; WOO, C. Interaction of Transonic Edge Dislocations with Self-interstitial Loop. *Radiat. Effects Defects Solids* **2010**, *157*, 201–208. [CrossRef]
18. Singh, B.; Golubov, S.; Trinkaus, H.; Serra, A.; Osetsky, Y.; Barashev, A. Aspects of microstructure evolution under cascade damage conditions. *J. Nucl. Mater.* **1997**, *251*, 107–122. [CrossRef]
19. Hull, D.; Bacon, D. *Introduction to Dislocations*, 5rd ed.; Elsevier Ltd.: Burlington, MA, USA, 2011; pp. 178–346.
20. Bonny, G.; Terentyev, D.; Elena, J.; Zinovev, A.; Minov, B.; Zhurkin, E. Assessment of hardening due to dislocation loops in bcc iron: Overview and analysis of atomistic simulations for edge dislocations. *J. Nucl. Mater.* **2016**, *473*, 283–289. [CrossRef]
21. Bacon, D.; Osetsky, Y. Dislocation-Obstacle Interactions at Atomic Level in Irradiated Metals. *Math. Mech. Solids* **2009**, *14*, 270–283. [CrossRef]
22. Zhu, B.; Huang, M.; Li, Z. Atomic level simulations of interaction between edge dislocations and irradiation induced ellipsoidal voids in alpha-iron. *Nucl. Instrum. Methods Phys. Res. B* **2017**, *397*, 51–61. [CrossRef]
23. Haghighat, S.; Fivel, C.; Fikar, J.; Schaeublin, R. Dislocation–void interaction in Fe: A comparison between molecular dynamics and dislocation dynamics. *J. Nucl. Mater.* **2009**, *386*, 102–105. [CrossRef]
24. Pelfort, M.; Osetsky, Y.; Serra, A. Vacancy interaction with glissile interstitial clusters in bcc metals. *Philos. Mag. Lett.* **2001**, *81*, 803–811. [CrossRef]
25. Puigvi, M.; Osetsky, Y.; Serra, A. Interactions between vacancy and glissile interstitial clusters in iron and copper. *Mater. Sci. Eng. A* **2004**, *365*, 101–106. [CrossRef]
26. Puigvi, M.; Serra, A.; Diego, N.; Osetsky, Y.; Bacon, D. Features of the interactions between a vacancy and interstitial loops in metals. *Philos. Mag. Lett.* **2004**, *84*, 257–266. [CrossRef]
27. Wang, S.; Guo, W.; Yuan, Y.; Zhu, X.; Cheng, L.; Cao, X.; Fu, E.; Shi, L.; Gao, F.; Lu, G. Evolution of vacancy defects in heavy ion irradiated tungsten exposed to helium plasma. *J. Nucl. Mater.* **2020**, *532*, 152051. [CrossRef]
28. Wang, H.; Zhang, C.; Yong, G.; Zeng, Z. Creeping Motion of Self Interstitial Atom Clusters in Tungsten. *Sci. Rep.* **2014**, *4*, 5096.
29. Zhou, W.; Li, Y.; Huang, L.; Zeng, Z.; Ju, X. Dynamical behaviors of self-interstitial atoms in tungsten. *J. Nucl. Mater.* **2013**, *437*, 438–444. [CrossRef]
30. Marinica, M.-C.; Ventelon, L.; Gilbert, M.; Proville, L.; Dudarev, S.; Marian, J.; Bencteux, G.; Willaime, F. Interatomic potentials for modelling radiation defects and dislocations in tungsten. *J. Phys. Condens. Matter* **2013**, *25*, 395502. [CrossRef]
31. Tomas, D.; Kazuto, A.; Hirotaro, M.; Hidehiro, Y.; Minoru, I.; Kouji, M.; Masahito, U.; Sergei, D. Fast, vacancy-free climb of prismatic dislocation loops in bcc metals. *Sci. Rep.* **2016**, *6*, 30596.
32. Yang, S. On The Elastic Interaction Between Dislocation Loop And Lattice Vacancy. *Acta Phys. Sin.* **1964**, *20*, 720–727. [CrossRef]

Article

A Mechanistic Study of Clustering and Diffusion of Molybdenum and Rhenium Atoms in Liquid Sodium

Zhixiao Liu [1,*], Mingyang Ma [2,*], Wenfeng Liang [2] and Huiqiu Deng [3]

1. College of Materials Science and Engineering, Hunan University, Changsha 410082, China
2. Institute of Nuclear Physics and Chemistry, China Academy of Engineering Physics, Mianyang 621900, China; liangwf@caep.cn
3. School of Physics and Electronics, Hunan University, Changsha 410082, China; hqdeng@hnu.edu.cn
* Correspondence: zxliu@hnu.edu.cn (Z.L.); mamingyang@caep.cn (M.M.)

Abstract: Liquid Na is widely used as the heat transfer medium in high-temperature heat pipes based on Mo-Re alloys. In this study, ab initio molecular dynamics are employed in order to understand the interactions between the Na solvent and Mo or Re solute in the liquid phase. Both the temperature and concentration effects on the clustering and diffusion behaviors of solute atoms are investigated. It is found that Mo_2 and Re_2 dimers can be stabilized in liquid Na, and the higher temperature leads to a stronger binding force. Pure Re and Mo-Re mixed solutes can form tetramers at the highest concentration. However, for the pure Mo solute, Mo_4 is not observed. The diffusivities of a single solute atom and clusters are calculated. It is found that the Mo species diffuse faster than the Re species, and the diffusivity decreases as the cluster size increases.

Keywords: clustering; diffusivity; Na solvent; Mo or Re solute; ab initio molecular dynamics

1. Introduction

In recent years, photovoltaic cells, electrochemical energy storage devices and wind turbines have been greatly improved to reduce the energy risk and environmental problems caused by the utilization of fossil fuels [1–4]. However, these sustainable energy technologies have not met the demand of the fast developed modern society. Going beyond the technologies discussed above, nuclear fission has been delivering green and reliable energy for half a century, and nuclear energy is a competitive candidate to mitigate energy risks [5–7]. In addition, nuclear energy is also expected to power spacecraft [8,9].

Heat pipes (HPs) are the key device in the nuclear energy system for achieving a high efficiency and safety. In nuclear reactors, the high-temperature HPs usually use liquid sodium (Na) as the heat transfer medium, because liquid Na possesses a high latent heat of vaporization, low saturated vapor pressure and outstanding heat conductivity [10]. Molybdenum (Mo) alloys with a low rhenium (Re) content can be used as structural materials in high-temperature HPs for nuclear application due to its high melting point, good mechanical properties and adequate irradiation resistance [11,12]. The Mo-based HPs accompanied with liquid Na working fluid can be operated at the temperature window of approximately 1000 K to 1700 K [13]. Comparing with other refractory metals, Mo also exhibits a relatively higher corrosion resistance to liquid alkali metal. Inoue et al. [14] reported that Mo alloys showed a better corrosion resistance than Nb alloys in a liquid Na environment. Saito et al. [15] studied the corrosion of niobium (Nb)-based alloys and Mo-based alloys, and found that the weight of the corrosion product of Mo alloys was ten times smaller than that of the Nb alloys in liquid lithium (Li). In addition, their results also demonstrated that metal elements in Nb alloys are dissolved more easily.

The dissolution, migration and precipitation of alloy elements in HPs can change the properties of the material surface, which is harmful for the performance of HPs [16–18]. For Mo-based HPs using Na as the working fluid, it is still not well understood how Mo and

Re atoms diffuse and accumulate in the Na solvent at the atomic scale. In the last few years, ab initio molecular dynamics (AIMD) have been successfully employed to investigate the properties of liquid alkali metal at extreme conditions [19–21]. In our present study, AIMD simulations are performed to reveal the interactions between liquid Na and Mo or Re solute atoms.

2. Computational Methods

All simulations in this study were performed on Vienna ab initio Simulation Package (VASP) [22,23] based on the density functional theory (DFT) [24,25]. The projector augmented-wave (PAW) method [26] was employed for describing the ion—electron interactions and the Perdew–Burke–Ernzerhof (PBE) functional [27] was used to describe the electron—electron exchange correlations. All AIMD simulations were carried out using NVT ensemble with the 400 eV energy cutoff of plane wave basis sets. AIMD simulations were run for at least 60 ps with a timestep of 2 fs. Only the Γ point was sampled in the first Brillouin zone.

The liquid metal model was constructed by randomly distributing Na atoms in a $15 \times 15 \times 15$ Å^3 box with a three-dimensional boundary condition. The number of Na atoms was determined by the liquid Na density reported by Argonne National Laboratory [28]. In this study, three temperature conditions of 700 K, 1100 K and 1600 K are considered, and corresponding number of Na atoms in the model are 75, 67 and 56, respectively, which correspond to the liquid Na density of 852 kg/m^3, 756 kg/m^3 and 626 kg/m^3, as reported in Ref. [28].

3. Results and Discussion

In this study, the pair correlation function $g(r)$ is generated by the VASPKIT code [29] in order to characterize the structure of the liquid metal system. The function $g(r)$ is formalized as

$$g(r) = \frac{1}{4\pi r^2} \frac{1}{N\rho} \sum_{i=1}^{N} \sum_{j \neq i}^{N} < \delta(r - |r_i - r_j|) > \quad (1)$$

where r is the radial distance, N is the total number of atoms in the modeling system and ρ is the average density of the system. The purpose of normalizing the $g(r)$ function by the density is to ensure that the radial distribution approaches unity for the long radial distance. Following Equation (1), the partial pair correlation function between two elements A and B can be written as

$$g_{AB}(r) = \frac{1}{4\pi r^2} \frac{N}{N_A N_B \rho} \sum_{i \in A}^{N} \sum_{j \in B, j \neq i}^{N} < \delta(r - |r_i - r_j|) > \quad (2)$$

Figure 1 shows the total pair correlation function $g(r)$ of liquid Na at 700 K, 1100 K and 1600 K after a 60 ps simulation. The primary peak of $g(r)$ is located between 3 Å and 4 Å, which agrees well with Li et al's AIMD results [21]. It is interesting that our present simulation shows that there is a small peak before the primary peak at a relatively low temperature, and that this small peak disappears at 1600 K. This small peak was also experimentally observed in the $g(r)$ function of liquid cesium when the temperature was below 400 K [30]. Figure 1 also demonstrates that the primary peak decreases and broadens obviously as the temperature increases. This phenomenon was also reported by Bickham and his collaborators [31].

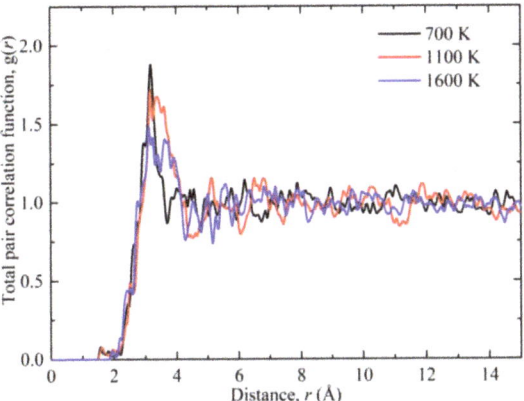

Figure 1. Total pair correlation function of liquid sodium at different temperature conditions.

The self-diffusion of Na atoms in the liquid phase is also evaluated. The diffusivity (D) can be calculated based on the Einstein–Stokes equation:

$$\lim_{t \to \infty} < R^2(t) > = 6Dt \tag{3}$$

here, $< R^2(t) >$ is the averaging mean-square displacement of Na. When recording the data, the system is equilibrated for 30 ps first, and then another 30 ps of simulation is run for collecting data.

According to the present study, the Na diffusivities are 13.2×10^{-5} cm^2/s, 21.4×10^{-5} cm^2/s and 32.6×10^{-5} cm^2/s at 700 K, respectively. According to the result of Li et al, the Na self-diffusion coefficient at 723 K was 12.8×10^{-5} cm^2/s [21], which is close to our present study. Based on experimental data, Meyer fitted an equation to predict the Na self-diffusion coefficient as [32]

$$D = 1.01 \times 10^{-3} \exp\left(-\frac{10,174}{RT}\right) \tag{4}$$

It is worth noting that, in Meyer's work, the experimental data were collected below the temperature of 500 K. According to Equation (4), the Na self-diffusion coefficient is 19.2 cm^{-2}/s at 700 K, which is higher than the present and previous AIMD results. However, all computational results yield a magnitude order of 10^{-5} cm^2/s.

The liquid Na system with Mo or Re atoms is modeled by replacing Na atoms with solute atoms. The effect of the solute concentration on the diffusion and clustering behaviors are investigated in the present study. Here, replacing one Na atom with a solute atom in the $15 \times 15 \times 15$ Å^3 computational domain corresponds to a concentration of 0.5 mol/L, replacing two Na atoms with solute atoms corresponds to 1 mol/L and replacing four Na atoms with solute atoms corresponds to 2 mol/L.

Figure 2 shows the partial pair correlation function of one solute atom in the computational domain (0.5 mol/L). It is found that the primary peak of the Mo-Na pair is lower than that of the Re-Na pair. Compared with the Re-Na pair, the primary peak position of the Mo-Na pair also shifts to the right. Therefore, it can be inferred that the solute Re atom can coordinate with more Na atoms than the solute Mo atom. The Bader charge analysis [33] is also performed in order to understand the interaction between the solute and the solvent from the aspect of the electronic structure. It is found that the net charge on the Mo atom is approximately $-2.2 \sim -3.0$ |e|, and the net charge on the Re atom is approximately $-2.8 \sim -3.4$ |e|. There is more negative charge transferred from the solvent atoms to the Re atom, which leads to the stronger Re-Na bonds.

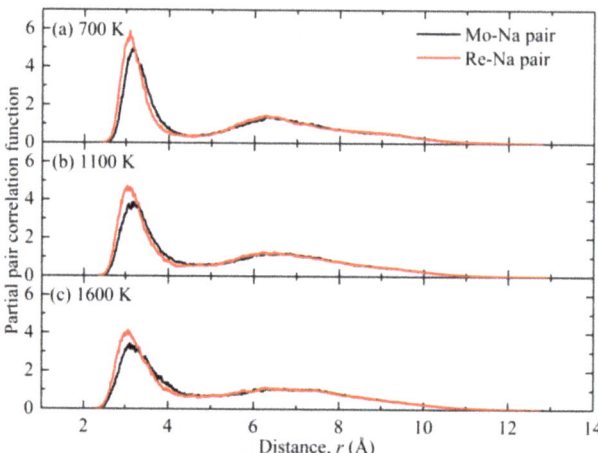

Figure 2. Partial pair correlation function of 0.5 mol/L solute atoms at the temperature of (**a**) 700 K, (**b**) 1100 K and (**c**) 1600 K.

At the relatively higher concentration, the clustering of solute atoms is observed. Snapshots in Figure 3 show the final structures of the two solute atoms in liquid Na at different temperatures after 60 ps AIMD simulations. It is interesting to find that the solute atoms can stabilize as a dimer in the liquid Na. Figure 4 demonstrates the Mo-Mo distance evolution during the AIMD simulation. The dimer is dynamically stable once it forms, and the average Mo-Mo bond length is 1.79 Å, 1.85 Å and 1.91 Å at 700 K, 1100 K and 1600 K, respectively. Here the Mo-Mo bond length is calculated based on the last 1000 AIMD steps. At 700 K, the dimer forms after 38 ps, and the increase in temperature can speed up the formation of the dimers. It should be noted that Figure 4 cannot reflect the dimerization rate because each case is only run once.

The thermodynamic stability of the dimer is also estimated by the binding energy, which is expressed as

$$E_b = 2\overline{E}_{M_1 Na_{N-1}} - (\overline{E}_{Na_N} + \overline{E}_{M_2 Na_{N-2}}) \tag{5}$$

Here, $\overline{E}_{Na_N}$ is the energy of the pure liquid Na without solute atoms, $\overline{E}_{M_1 Na_{N-1}}$ is the liquid Na system with one solute atom, $\overline{E}_{M_2 Na_{N-2}}$ is the energy of the system with two solute atoms and the subscript "N" represents the number of atoms in the model. All energetic terms are the average value of the last 1000 simulation steps. Table 1 shows the binding energy of dimers at different temperature conditions. The positive binding energy indicates that forming dimers is an exothermic reaction and that dimers are thermodynamically stable at the temperature range of 700 K to 1600 K.

Table 1. Biding energy of Mo_2 and Re_2 dimers at different temperature.

Temperature (K)	Binding Energy, E_b (eV)	
	Mo_2	Re_2
700	5.46	3.83
1100	7.35	4.53
1600	7.95	5.29

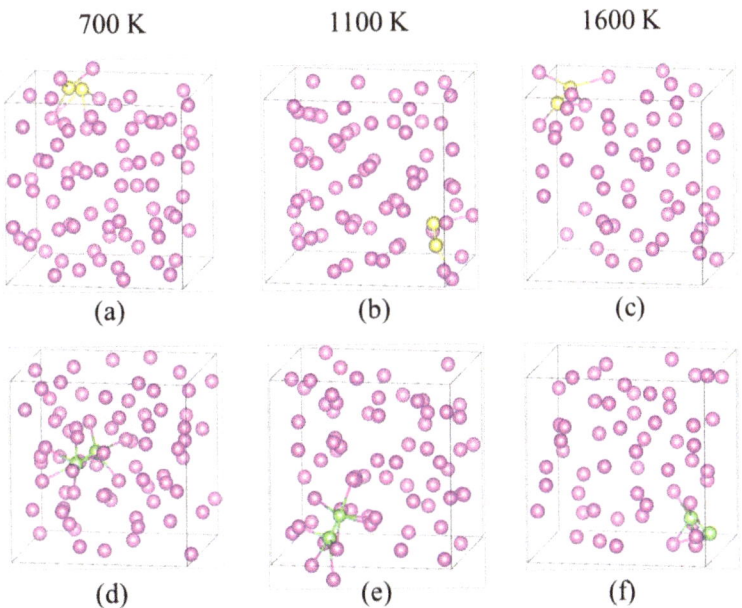

Figure 3. Snapshots of two solute atoms (1 mol/L) in liquid Na at different temperatures after 60 ps AIMD simulation: (**a**–**c**) Mo atoms in liquid Na at 700 K, 1100 K and 1600 K, respectively; (**d**–**f**) Re atoms in liquid Na at 700 K, 1100 K and 1600 K, respectively. Violet spheres, yellow spheres and green spheres represent Na atoms, Mo atoms and Re atoms, respectively.

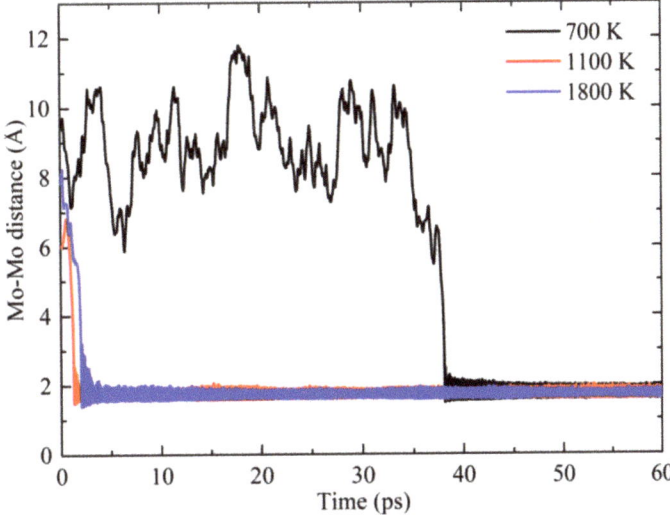

Figure 4. Mo-Mo distance evolution for two Mo atoms in the liquid Na system at different temperatures.

It is worth noting that the binding energy increases as the temperature increases. The Bader charge analysis is performed to calculate the net charge on dimers shown in Figure 5. The net charge on Mo_2 is −3.6 |e| at 700 K, −2.4 |e| at 1100 K and −2.1 |e| at 1600 K. As discussed in Figure 2, the solute atom captures electrons from surrounding Na atoms, and fewer Na atoms coordinate with the solute atom when increasing the temperature. Therefore, the Mo_2 dimer at a higher temperature gains less of a net charge. It is worth noting that the binding energy of the dimer is proportional to the net charge (Figure 5). For the Mo_2 molecule, the lowest unoccupied molecular orbital (LUMO) is empty doubly degenerate δ_{4d}^* antibonding orbitals. The migration of electrons from the Na atoms to the Mo_2 dimer will occupy these antibonding orbitals. As the occupation of the antibonding orbitals increases, the Mo-Mo interaction weakens, which leads to a lowering of the binding energy. The net charge of the Re_2 dimer is -4.1 |e| at 700 K, -3.9 |e| at 1100 K and -2.3 |e| at 1600 K. As with the Mo_2 dimer, the binding energy of the Re_2 dimer is also proportional to the net charge. It is worth noting that Re 5d orbitals have one more electron than Mo 4d orbitals. When forming a neutral Re_2 dimer, its δ_{5d}^* antibonding orbitals are halfly occupied, which indicates that the bond of neutral Mo_2 is weaker than Re_2. In addition, the Re_2 dimer can capture more electrons from the Na solvent than the Mo_2 dimer. Hence, the occupation of the antibonding orbitals of Re_2 in the liquid Na is much higher than Mo_2, which results in the much lower binding energy. Taking Re_2 in liquid Na at 700 K, for instance, the net charge on this diatomic molecule is -4.1 |e|. In this case, the antibonding δ_{5d}^* orbitals are fully occupied, and the antibonding π_{5d}^* orbitals are also halfly occupied, which lead to the lowest binding energy of 3.83 eV.

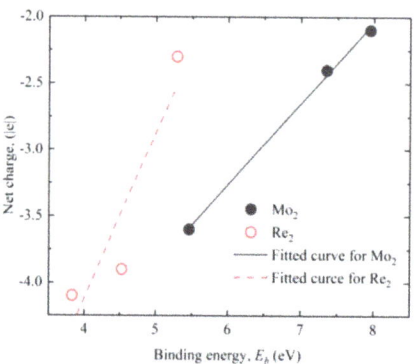

Figure 5. The relationship between the binding energy of the dimer and the net charge.

The clustering behaviors of four solute atoms (corresponding to the concentration of 2 mol/L) in the liquid Na are investigated. The solute atoms can be four Mo atoms, four Re atoms or a mixture of two Mo atoms and two Re atoms. Figure 6 shows the total pair correlation function of the liquid system after a 60 ps simulation. For models including four Mo atoms, they still exhibit the typical liquid characterization, as shown in Figure 1. However, obvious peaks located around 2 Å can be found in Figure 6e–i. These peaks can be attributed to the segregation of solute atoms [21]. The atomic configurations of liquid systems after the 60 ps AIMD simulation are shown in Figure 7. It is interesting that the Mo solute and Re solute exhibit relatively different behaviors. Four Mo atoms form two Mo_2 dimers, but cannot form a stable Mo_4 tetramer. It should be noticed that these two dimers interact with each other and diffuse together (see videos in the Supplementary Material).

For the Re solute, Re_4 is observed at 1100 K and 1600 K after a 60 ps simulation. Li et al. also found a cerium tetramer (Ce_4) in the liquid Na using the AIMD simulation [21]. In their study, forming Ce_4 was attributed to the rejection of the liquid phase to the solute atoms. However, our simulation has demonstrated that there are attractive interactions between solvent Na atoms and Re atoms. It is known that the outmost shell of a Na atom is 3s orbital, whereas the outmost shell of a Re atom is 5d orbital. The energy level of the former is much lower than the latter. According to the linear combination of the atomic orbitals–molecular orbitals (LCAO-MO) theory, molecular orbitals are always formed by atomic orbitals with a small energy difference. Therefore, the Re-Re attractive interaction is stronger than that of the Re-Mo pair, which leads to the forming of the large Re cluster in the liquid Na. For the mixed solute, the formation of Mo_2Re_2 is also observed, as shown in Figures 6g–i and 7g–i. The spin states of dimers and tetramers are also investigated. It is found that the Mo species does not show spin polarization. For the Re species, the magnetic momentum of Re2 is around 0.8 μ_b per Re atom, whereas Re4 does not show an obvious spin polarization.

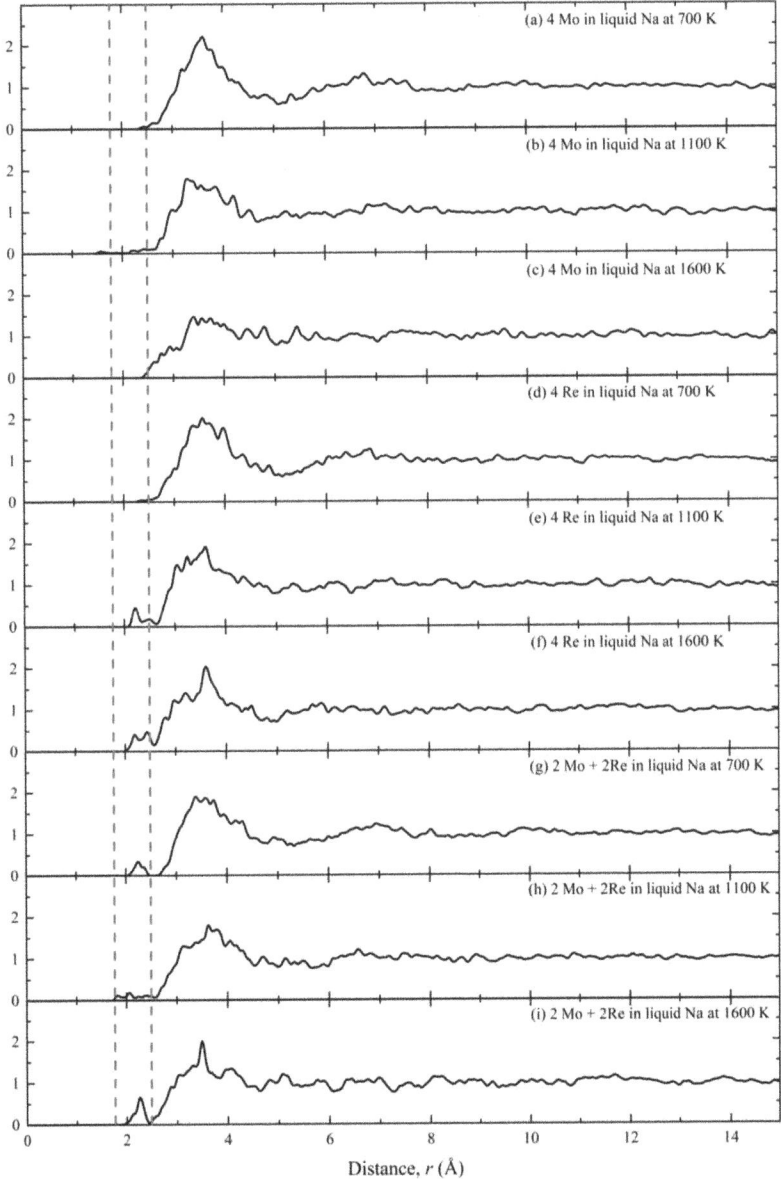

Figure 6. Total pair correlation function of four solute atoms in liquid Na at different systems: (**a–c**) four Mo atoms in liquid Na at 700 K, 1100 K and 1600 K, respectively; (**d–f**) four Re atoms in liquid at different temperatures; (**g–i**) two Mo atoms and two Re atoms in liquid Na at different temperatures.

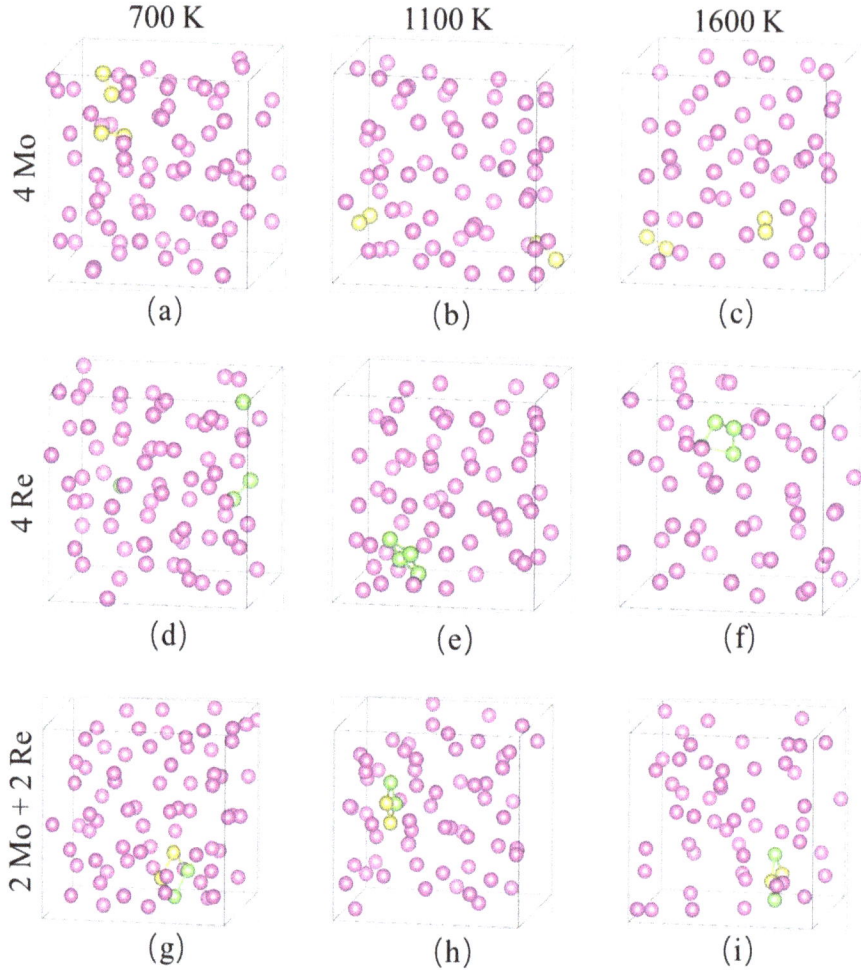

Figure 7. Structures of four solute atoms in liquid Na after 60 ps AIMD simulation: (**a**–**c**) four Mo atoms in liquid Na at 700 K, 1100 K and 1600 K; (**d**–**e**) four Re atoms in liquid Na at different temperatures; (**g**–**i**) two Mo atoms and two Re atoms in liquid Na at different temperatures.

In the present work, the accumulation behavior of six Mo or Re atoms is investigated at 1600 K. It is found that Mo atoms will form three interacting Mo_2 dimers, whereas Re atoms form an octahedral-like Re_6 cluster.

Understanding mass transport is essential for evaluating the performance and service life of alkali metal high-temperature heat pipes. In the present study, the diffusivities of single solute atoms and small clusters are calculated and shown in Figure 8. As expected, the diffusivity of any species is dependent on the temperature, and a higher temperature leads to a faster diffusion. In addition, the increase in cluster size can reduce the diffusivity. The diffusivities of the Mo_x (x = 1, 2, 4) clusters are approximately two times as high as those of the Re_x clusters, because the atomic weight of Mo (95.94 a. u.) is only half of the atomic weight of Re (186.2 a. u.). The diffusivity of a single Re atom is even lower than the Mo_2 cluster. As discussed above, there are not any stable Mo_4 clusters formed during the AIMD simulation, but two Mo_2 clusters can interact with each other and move together. The interaction between the two Mo_2 dimers can enhance the diffusion effects at low-temperature conditions, as shown in Figure 8. At 700 K, the diffusivity of one Mo_2 dimer is 2.53×10^{-5} cm^2/s, whereas the diffusivity of two interacting Mo_2 dimers is 2.81×10^{-5} cm^2/s. At 1100 K, the diffusivity of two interacting Mo_2 dimers is also approximately 0.3×10^{-5} cm^2/s higher than that of a Mo_2 dimer.

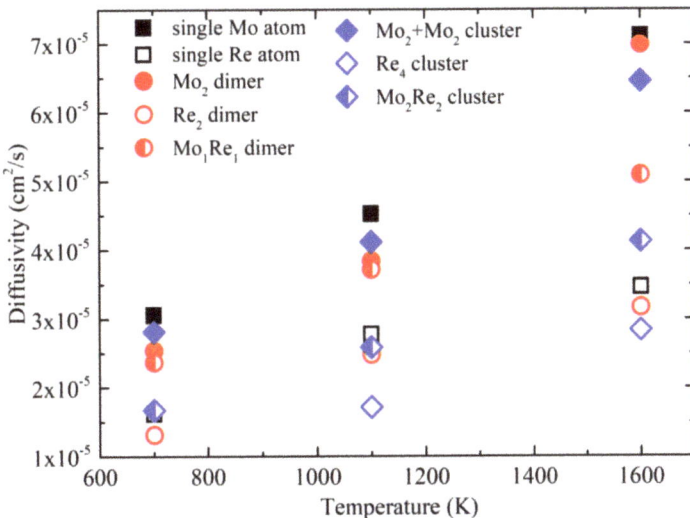

Figure 8. Diffusivity of Mo and Re clusters in liquid Na at different temperatures.

4. Conclusions

In this study, AIMD simulations are performed in order to understand the clustering and diffusion behaviors of Mo and Re atoms in liquid Na. For a single solute atom, the Re-Na interaction is stronger than the Mo-Na interaction. For models with two solute atoms, both Mo_2 dimers and Re_2 dimers are found in the Na solvent. The binding energy of the dimer is proportional to the occupation of the antibonding orbitals of the dimer. The higher occupation always leads to a lower binding energy. For models with four solute atoms, four Mo atoms form two interacting Mo_2 dimers, but a stable Mo_4 tetramer is not observed. However, for the pure Re solute or mixed solute, a stable tetramer is observed. The diffusivity of the Mo species is much faster than that of the Re species, and the diffusivity decreases as the cluster size increases.

Supplementary Materials: The following are available online at https://www.mdpi.com/article/10.3390/met11091430/s1, Video S1: Trajectory of four Mo atoms in Na at 1600 K; Video S2: Trajectory of four Re atoms in Na at 1600 K.

Author Contributions: Z.L. and M.M. concept this work; Z.L. performs simulations, collects data and writes the original draft; all authors analyze results and polish the manuscript. W.L. and H.D. have contributed writing–review and editing manuscript. All authors have read and agreed to the published version of the manuscript.

Funding: The President Foundation of China Academy of Engineering Physics (No. YZJJLX2018002); Fundamental Research Funds for the Central Universities.

Institutional Review Board Statement: Not applicable.

Informed Consent Statement: Not applicable.

Data Availability Statement: Not applicable.

Acknowledgments: This work is financially supported by the President Foundation of China Academy of Engineering Physics (No. YZJJLX2018002). Z. Liu also thanks Fundamental Research Funds for the Central Universities.

Conflicts of Interest: The authors declare no conflict of interest.

References

1. Tachan, Z.; Rühle, S.; Zaban, A. Dye-sensitized solar tubes: A new solar cell design for efficient current collection and im-proved cell sealing. *Sol. Energy Mater. Sol. Cells* **2010**, *94*, 317–322. [CrossRef]
2. Zhou, L. Progress and problems in hydrogen storage methods. *Renew. Sustain. Energy Rev.* **2005**, *9*, 395–408. [CrossRef]
3. Kim, T.; Song, W.; Son, D.-Y.; Ono, L.K.; Qi, Y. Lithium-ion batteries: Outlook on present, future, and hybridized technologies. *J. Mater. Chem. A* **2019**, *7*, 2942–2964. [CrossRef]
4. Foley, A.M.; Leahy, P.G.; Marvuglia, A.; Mckeogh, E.J. Current methods and advances in forecasting of wind power genera-tion. *Renew. Energy* **2012**, *37*, 1–8. [CrossRef]
5. Ion, S. Nuclear energy: Current situation and prospects to 2020, Philosophical Transactions of the Royal Society a-Mathematical. *Phys. Eng. Sci.* **2007**, *365*, 935–944.
6. Nifenecker, H. Future electricity production methods. Part 1: Nuclear energy. *Rep. Prog. Phys.* **2011**, *74*, 022801. [CrossRef]
7. Pravalie, R.; Bandoc, G. Nuclear energy: Between global electricity demand, worldwide decarbonisation imperativeness, and planetary environmental implications. *J. Environ. Manag.* **2018**, *209*, 81–92. [CrossRef]
8. El-Genk, M.S. Deployment history and design considerations for space reactor power systems. *Acta Astronaut.* **2009**, *64*, 833–849. [CrossRef]
9. Li, Z.; Yang, X.; Wang, J.; Zhang, Z. Off-design performance and control characteristics of space reactor closed Brayton cycle system. *Ann. Nucl. Energy* **2019**, *128*, 318–329. [CrossRef]
10. Grover, G.M.; Cotter, T.P.; Erickson, G.F. Structures of Very High Thermal Conductance. *J. Appl. Phys.* **1964**, *35*, 1990–1991. [CrossRef]
11. Leonhardt, T.; Carlén, J.; Buck, M.; Brinkman, C.R.; Stevens, C.O. Investigation of mechanical properties and microstructure of various molybdenum-rhenium alloys. *AIP Conf. Proc.* **1999**, *458*, 685–690. [CrossRef]
12. Busby, J.; Leonard, K.; Zinkle, S. Radiation-damage in molybdenum–rhenium alloys for space reactor applications. *J. Nucl. Mater.* **2007**, *366*, 388–406. [CrossRef]
13. El-Genk, M.S.; Tournier, J.-M. A review of refractory metal alloys and mechanically alloyed-oxide dispersion strengthened steels for space nuclear power systems. *J. Nucl. Mater.* **2005**, *340*, 93–112. [CrossRef]
14. Inoue, S.; Kano, S.; Saito, J.-I.; Isshiki, Y.; Yoshida, E.; Morinaga, M. Corrosion Behaviour of Nb-Based and Mo-Based Alloys in Liquid Na. In *Liquid Metal Systems*; Springer: Berlin/Heidelberg, Germany, 1995; pp. 75–83. [CrossRef]
15. Saito, J.-I.; Inoue, S.; Kano, S.; Yuzawa, T.; Furui, M.; Morinaga, M. Alloying effects on the corrosion behavior of binary Nb-based and Mo-based alloys in liquid Li. *J. Nucl. Mater.* **1999**, *264*, 216–227. [CrossRef]
16. Tu, S.-T.; Zhang, H.; Zhou, W.-W. Corrosion failures of high temperature heat pipes. *Eng. Fail. Anal.* **1999**, *6*, 363–370. [CrossRef]
17. Yoshida, E.; Furukawa, T. Corrosion issues in sodium-cooled fast reactor (SFR) systems. In *Nuclear Corrosion Science and Engineering*; Woodhead Publishing: Sawston, UK, 2012; pp. 773–806. [CrossRef]
18. Brissonneau, L. New considerations on the kinetics of mass transfer in sodium fast reactors: An attempt to consider irradia-tion effects and low temperature corrosion. *J. Nucl. Mater.* **2012**, *423*, 67–78. [CrossRef]
19. Koci, L.; Ahuja, R.; Vitos, L.; Pinsook, U. Melting of Na at high pressure from ab initio calculations. *Phys. Rev. B* **2008**, *77*, 132101. [CrossRef]
20. Yuryev, A.A.; Gelchinski, B.R. Simulation of properties of liquid alkali metals at high temperatures and pressures by ab initio molecular dynamics method. *Dokl. Phys.* **2015**, *60*, 105–108. [CrossRef]
21. Li, X.; Samin, A.; Zhang, J.; Unal, C.; Mariani, R. Ab-initio molecular dynamics study of lanthanides in liquid sodium. *J. Nucl. Mater.* **2016**, *484*, 98–102. [CrossRef]
22. Kresse, G.; Furthmüller, J. Efficiency of ab-initio total energy calculations for metals and semiconductors using a plane-wave basis set. *Comput. Mater. Sci.* **1996**, *6*, 15–50. [CrossRef]
23. Kresse, G.; Hafner, J. Ab initio. *Phys. Rev. B* **1993**, *48*, 13115–13118. [CrossRef]
24. Car, R.; Parrinello, M. Unified Approach for Molecular Dynamics and Density-Functional Theory. *Phys. Rev. Lett.* **1985**, *55*, 2471–2474. [CrossRef]
25. Jones, R.O.; Gunnarsson, O. The density functional formalism, its applications and prospects. *Rev. Mod. Phys.* **1989**, *61*, 689–746. [CrossRef]
26. Blöchl, P.E. Projector augmented-wave method. *Phys. Rev. B* **1994**, *50*, 17953–17979. [CrossRef] [PubMed]
27. Perdew, J.P.; Burke, K.; Ernzerhof, M. Generalized Gradient Approximation Made Simple. *Phys. Rev. Lett.* **1996**, *77*, 3865–3868. [CrossRef] [PubMed]
28. Fink, J.; Leibowitz, L. *Thermodynamic and Transport Properties of Sodium Liquid and Vapor*; Argonne National Laboratory: Argonne, IL, USA, 1995. [CrossRef]
29. Wang, V.; Xu, N.; Liu, J.-C.; Tang, G.; Geng, W.-T. VASPKIT: A user-friendly interface facilitating high-throughput computing and analysis using VASP code. *Comput. Phys. Commun.* **2021**, *267*, 108033. [CrossRef]

30. Winter, R.; Hensel, F.; Bodensteiner, T.; Gläser, W. The static structure factor of cesium over the whole liquid range up to the critical point. *Z. Elektrochem. Ber. Bunsenges. Phys. Chem.* **2015**, *91*, 1327–1330. [CrossRef]
31. Bickham, S.R.; Pfaffenzeller, O.; Collins, L.A.; Kress, J.D.; Hohl, D. Ab initio molecular dynamics of expanded liquid sodium. *Phys. Rev. B* **1998**, *58*, R11813–R11816. [CrossRef]
32. Meyer, R.E.; Nachtrieb, N.H. Self-Diffusion of Liquid Sodium. *J. Chem. Phys.* **1955**, *23*, 1851–1854. [CrossRef]
33. Tang, W.; Sanville, E.; Henkelman, G. A grid-based Bader analysis algorithm without lattice bias. *J. Phys. Condens. Matter Inst. Phys. J.* **2009**, *21*, 084204. [CrossRef]

Article

A First-Principles Study on Na and O Adsorption Behaviors on Mo (110) Surface

Qingqing Zeng [1], Zhixiao Liu [1,*], Wenfeng Liang [2], Mingyang Ma [2,*] and Huiqiu Deng [3]

1. College of Materials Science and Engineering, Hunan University, Changsha 410082, China; Qqzeng@hnu.edu.cn
2. Institute of Nuclear Physics and Chemistry, China Academy of Engineering Physics, Mianyang 621900, China; liangwf@caep.cn
3. School of Physics and Electronics, Hunan University, Changsha 410082, China; hqdeng@hnu.edu.cn
* Correspondence: zxliu@hnu.edu.cn (Z.L.); mamingyang@caep.cn (M.M.)

Abstract: Molybdenum-rhenium alloys are usually used as the wall materials for high-temperature heat pipes using liquid sodium as heat-transfer medium. The corrosion of Mo in liquid Na is a key challenge for heat pipes. In addition, oxygen impurity also plays an important role in affecting the alloy resistance to Na liquid. In this article, the adsorption and diffusion behaviors of Na atom on Mo (110) surface are theoretically studied using first-principles approach, and the effects of alloy Re and impurity O atoms are investigated. The result shows that the Re alloy atom can strengthen the attractive interactions between Na/O and the Mo substrate, and the existence of Na or O atom on the Mo surface can slower down the Na diffusion by increasing diffusion barrier. The surface vacancy formation energy is also calculated. For the Mo (110) surface, the Na/O co-adsorption can lead to a low vacancy formation energy of 0.47 eV, which indicates the dissolution of Mo is a potential corrosion mechanism in the liquid Na environment with O impurities. It is worth noting that Re substitution atom can protect the Mo surface by increasing the vacancy formation energy to 1.06 eV.

Keywords: Mo-Re alloy; Na adsorption and diffusion; surface vacancy; first-principles calculation

1. Introduction

Alkali metal heat pipes (HPs) are initially designed for heat transfer in space nuclear power systems, of which the operating temperature is typically from 800 K to 1800 K. HPs using alkali metals are also promising in advanced energy and power systems such as high-efficiency waste heat utilization [1], hypersonic vehicles [2], and molten salt reactors [3].

A heat pipe consists of a sealed shell, wick structure and a vapor chamber containing working fluid, which is normally filled after the shell is evacuated [4]. Heat transfer in a heat pipe is achieved passively by the phase change and the circulation of the working fluid [5]. Different types of working fluid and shell material are adopted in heat pipes used under different working conditions. The type of heat pipe can be divided into four main types according to their working temperature: low temperature heat pipe (−270~0 °C), normal temperature heat pipe (0~200 °C), medium temperature heat pipe (200~600 °C) and high temperature heat pipe (above 600 °C) [6].

Alkali metal heat pipes operating at temperature above 800 K have typically been constructed taking liquid alkali metal: potassium, sodium, or lithium as the working fluid due to their high-power capacity and great thermal stability [7]. For alkali metal heat pipes used in nuclear power systems, a key requirement is the compatibility of structure materials with both nuclear fuel and alkali metal [8]. Refractory metals and alloys, for owning both high creep strength at high temperatures and excellent compatibility with alkali liquid metals as well as nuclear fuel, are often applied as the shell material of alkali metal heat pipes [9]. Molybdenum (Mo) is a kind of refractory metal that is reported to be one of the greatest candidates for use of alkali metal heat pipe walls [10–13]. Recently, Mo alloys are also considered as structural materials using in nuclear reactors [14,15]. However, pure Mo becomes brittle at about room temperature and below, which impacts heavily not only on the fabrication of heat pipes, but also the transportation process that easily leading to heat pipe breakage [16,17]. To tackle this problem, adding rhenium (Re) into pure molybdenum is found to be effective in enhancing low temperature ductility while also improving the high-temperature strength and creep resistance, known as the "rhenium effect" [18,19]. Mo-Re alloy has great advantage in high temperature heat pipes, where the operating environment is usually highly oxidizing and corrosive.

One issue for Mo-Re alloy in the heat pipe is the corrosion induced by the liquid metal working fluid and impurities. Studies have shown that dissolution, mass transfer and impurity reactions are the major corrosion mechanisms in refractory metal-alkali systems. Meanwhile, the most serious corrosion problems encountered are related to impurity reactions associated with oxygen [20]. Even though the addition of Re can improve the low temperature ductility, creep resistance and high temperature strength, it is not known that whether the Re atom could bring an improvement on the corrosion resistance of Mo. In addition, as the existence of oxygen (O) could lead to serious corrosion problems, it is important to learn the chemical interaction between O and Mo surface [21]. In this study, we used a first-principles approach to investigate basic properties such as adsorption, diffusion properties of Na atom and O atom on the pure Mo (110) surface and Mo-Re (110) surface. In addition, the formation of surface vacancy was calculated for evaluating if the adsorbates can prevent Mo from dissolution.

2. Computational Methods and Model

We employed the Vienna Ab-initio Simulation Package (VASP, version 5.4, developed by Vienna University) [22,23] to carry out our first-principles calculations. All calculations were implemented based on density-functional theory (DFT) [24,25], using a plane wave basis set [26]. The interactions between the core and valence electrons were described with the projector augmented wave (PAW) approach [27,28]. Generalized gradient approximation (GGA) of the Perdew–Burke–Ernzerhof (PBE) functional [29] was adopted in all the calculations. An energy cut-off for the plane-wave basis set was set to 380 eV for both the relaxed and static computations. The residual force for structure optimization was less than 0.02 eV/Å. The $7 \times 7 \times 1$ k-point grids generated by the Monkhorst–Pack (MP) technique [30] was found to be sufficient for the present study.

The Mo (110) surface was used to explore the interaction between Na/O atoms and the Mo substrate because the Mo (110) surface has the lowest surface energy as reported in Materials Project Database [31]. Mo (110)-(2×2) surface cell was represented by a 5-layers slab model (8 atoms per layer) with the vacuum thickness of 10 Å. The upper two layers were fully relaxed and the bottom three layers were fixed as the bulk phase during structure optimization. In this study, we also investigated the effect of alloy element Re on the adsorption of Na and O atoms. For Mo-Re alloy, when the weight percentage of Re is less than 14%, Re atoms randomly replace Mo atoms at the lattice site [32]. We considered three adsorption sites for the Na and O atom on Mo/Mo-Re alloy: TOP site, Bridge site and Hollow site of Mo/Mo-Re (110) surface. Top site is located over the Mo atoms of the topmost layer, and bridge site is located between the two Mo atoms of the topmost layer, while Hollow site is above the Mo atoms of the second layer, as is shown in Figure 1. For one atom adsorbed on the (110)-(2×1) surface cell, the corresponding coverage (Θ) is $\frac{1}{8}$ ML.

The climbing image nudged-elastic-band (CI-NEB) method was employed to calculate the Na diffusion barriers between two most stable adsorption sites on Mo/Mo-Re (110) surface [33].

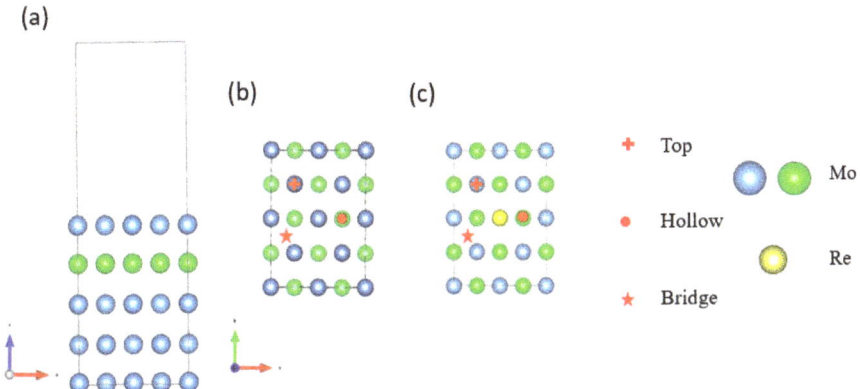

Figure 1. Structure of Mo/Mo-Re (110) surfaces and adsorption sites: (**a**) side view of Mo/Mo-Re (110) slab model; (**b**) Top view of Mo (110) surface and (**c**) top view of Mo-Re surface. Blue and green spheres represent Mo atoms, and yellow spheres represent Re atoms.

3. Results and Discussion

3.1. Model Verification

Molybdenum (Mo) has a body centered cubic (bcc) structure with an experimental lattice constant of 3.14 Å [34] and each conventional cell has 2 atoms. The theoretical lattice constant found in this study is 3.15 Å which agrees well with the experimental value.

In order to find out the thermodynamically most stable surface structure of bcc-Mo, the surface energy is calculated by the following equation:

$$\sigma = (E_{slab} - n \times E_{bulk})/(2A) \tag{1}$$

where A is the surface area and n is the number of atoms in the slab model, E_{slab} is the total energy of the surface, E_{bulk} is the energy per Mo atom of the ground state structure.

The surface energies of low-index Mo surfaces are shown in Table 1. The surface energies of Mo (100), Mo (110) and Mo (111) are 0.22 eV/Å^2, 0.17 eV/Å^2 and 0.18 eV/Å^2, respectively. Our results are the same with date reported in Materials Project Database [31]. All above results demonstrate that the input parameters and surface models used in the present study are reliable. In addition, considering that Mo (110) surface has the lowest surface energy, we focus on understanding Na and O adsorption properties on this energetically most stable surface.

Table 1. Surface energies of low-index surfaces.

	Surface Energy(eV/Å^2)	
Surface	Present Calculation	Materials Project Database
(100)	0.22	0.22
(110)	0.17	0.17
(111)	0.18	0.18

3.2. Na Atom and O Atom Adsorption

The adsorption behaviors of Sodium atom on both Mo (110) and Mo-Re (110) surfaces are investigated firstly. In the current calculations, the adsorption energy for the Na atom adsorbed on the Mo (110) or Mo-Re (110) surfaces is defined as:

$$E_{ads} = E_{(Mo/Mo-Re)-Na} - E_{(Mo/Mo-Re)slab} - E_{Na} \tag{2}$$

Here $E_{(Mo/Mo-Re)-Na}$ is the total energy of the adsorbate–substrate system, $E_{(Mo/Mo-Re)slab}$ is the total energy of the clean Mo/Mo-Re surface, and E_{Na} is the average energy of BCC Na. A negative E_{ads} value indicates the attractive interaction between the adsorbate and the substrate. The more negative adsorption energy implies the stronger attractive interaction between the adsorbate and the substrate.

Figure 2 shows the Initial states and final states of Na adsorption Mo (110) and Mo-Re (110) surfaces. The adsorption energies and structural parameters of Na on Mo or Mo-Re (110) surfaces are given in Table 2. For the Mo (110) surface, the hollow site is the energetically most favored for Na adsorption and the corresponding energy is −0.51 eV. The Bridge site is second preferred for Na adsorption with the adsorption energy of −0.44 eV. The Top site is a stable one for Na adsorption, but the adsorption energy is only −0.35 eV. For Mo-Re (110) surface, all un-equivalent Top, Hollow and Bridge sites are considered for Na adsorption. For each type of adsorption site, only the initial and final sites with the lowest energy are collected and shown in Figure 2 and Table 2. Compared with the pristine Mo (110) surface, Mo-Re (110) surface provides stronger affinity to the Na atom. It is worth noting that the Na atom initially placed at the Bridge site moves towards the Re atom after the structure optimization, as can be seen in Figure 2f, and this configuration delivers the lowest adsorption energy of −0.60 eV. The adsorption energies of Na adsorption on the Top site and Hollow site of Mo-Re (110) surface are −0.43 eV and −0.56 eV, respectively. The average vertical distance between adsorbed Na atom and the top layer (d_{Na-sur}) is also listed in Table 2. It is interesting to found that the Na adsorption energy is proportional to $d_{Na-surf}$ (Figure 3). A shorter distance between the Na atom and the top layer indicates a stronger attractive interaction. Both energetical and geometric parameters indicate that the Re atom can strengthen the attractive interaction between Na and the Mo surface. The electronic structure is analyzed for understanding the effect of Re atom on Na adsorption. On the Mo (110) surface, the Na atom only interacts with Mo atoms in the upmost layer, as shown in Figure 4a. The Re atom can significantly affect the charge redistribution induced a by Na adsorption. As shown in Figure 4b, electrons from the Mo settled in the second atomic layer migrate to the first layer, resulting to great electron accumulation. In this case, the columbic attraction between the Na atom and the substrate is enhanced and lead to a lower adsorption energy.

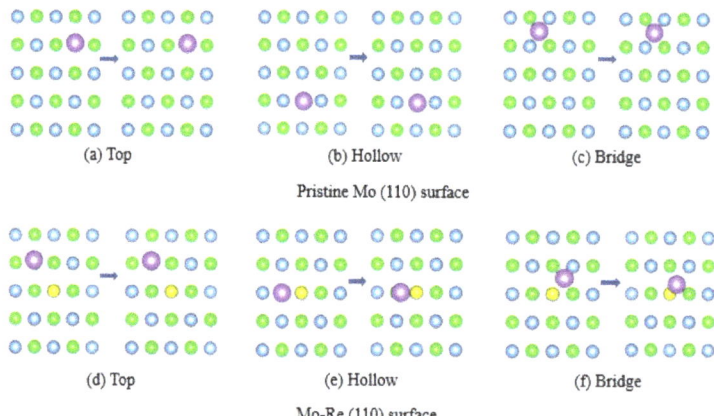

Figure 2. Initial and final configurations of Na adsorption on Mo (110) and Mo-Re (110) surfaces. Snaps (**a**–**c**) show Na atoms initially placed at Top site, Hollow site and Bridge site of Mo (110) surface. Snaps (**d**–**f**) show Na atoms initially placed at Top site, Hollow site and Bridge site of Mo-Re (110) surface. Blue and green spheres represent Mo atoms in the first layer and the second layer. Yellow and purple spheres represent Re atoms and Na atoms, respectively.

Table 2. Adsorption energy of Na on Mo (110) and Mo-Re (110) surfaces. E_{ads} is the adsorption energy, and $d_{Na-surf}$ is the vertical distance between the adsorbate and the top layer of the slab model.

Surface	Initial Site	d_{Na-sur} (Å)	E_{ads} (eV)	Final Site
Mo (110)	Top	2.92	−0.35	Top
	Hollow	2.74	−0.51	Hollow
	Bridge	2.82	−0.44	Bridge
Mo-Re (110)	Top	2.82	−0.43	Top
	Hollow	2.71	−0.56	Hollow
	Bridge	2.68	−0.60	Bridge

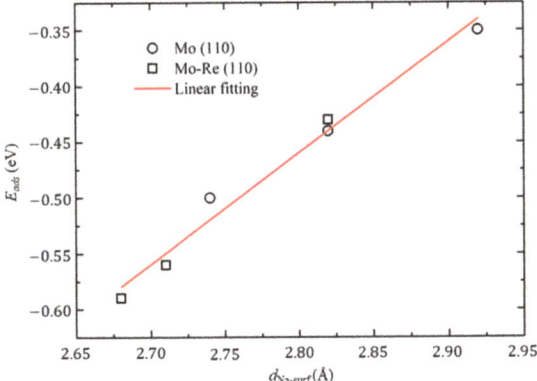

Figure 3. Na adsorption energy as the function of the average distance between the Na atom and the top layer of the substrate.

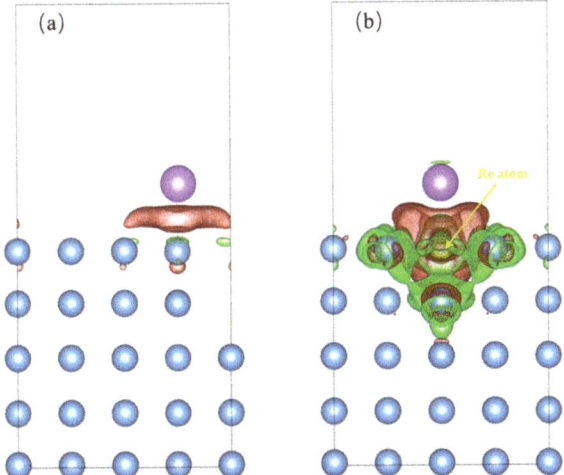

Figure 4. Difference charge density of Na adsorption at the hollow site of (a) Mo (110) surface and (b) Mo-Re (110) surface. The red isosurface represent the electron accumulation region, while the green isosurface represent the electron depletion region. The Re atom is located in the center of Mo (110) surface, as shown in Figure 1c.

The adsorption energies of O on Mo and Mo-Re (110) surfaces are also investigated in the present study (Table 3). The adsorption energy for the O atom adsorbed on the Mo/Mo-Re (110) surfaces is defined as

$$E_{ads} = E_{(Mo/Mo-Re)-O} - E_{(Mo/Mo-Re)slab} - \frac{1}{2}E_{O_2} \quad (3)$$

where $E_{(Mo/Mo-Re)-O}$ is the total energy of the adsorbate–substrate system, $E_{(Mo/Mo-Re)slab}$ is the total energy of clean Mo/Mo-Re surfaces, and E_{O_2} is the energy of an isolated O_2 molecule. The adsorption energy of a single O atom on the surface is always calculated to characterize the oxygen-substrate interactions for refractory materials, and the energy of an O atom is usually referenced to the half of the O_2 molecule [35–37].

Table 3. Adsorption energy of O on Mo/Mo-Re (110) surface.

Surface	Initial Site	d_{O-surf} (Å)	E_{ads} (eV)	Final Site
Mo (110)	Top	2.18	−2.80	Top
	Hollow	1.15	−4.09	Hollow
	Bridge	1.15	−4.09	Hollow
Mo-Re (110)	Top	1.73	−2.82	Top
	Hollow	1.14	−4.14	Hollow
	Bridge	1.14	−4.14	Hollow

As listed in Table 3, The Hollow site is energetically preferred for O adsorption on the Mo (110) surface with the adsorption energy of −4.09 eV. It is worth noting that the O atom initially placed at the Bridge site will spontaneously move to the Hollow site after the structure optimization as shown in Figure 5c. The O atom can also be stabilized at the Top site, but the adsorption energy is only −2.80 eV. As with Na on Mo (110) surface, the shorter vertical distance between O atom and the substrate (d_{O-surf}) leads to a lower (more negative) adsorption energy. For the Mo-Re (110) surface, the Hollow site is also the most favored for O adsorption and the corresponding adsorption energy is −4.14 eV, which is even 0.05 lower than the adsorption energy of O at the Hollow site of Mo (110) surface. It is worth mentioning that all un-equivalent Hollow sites around the Re atoms are checked, and Figure 5 as well as Table 3 demonstrates configurations with the lowest energy. As with the pristine Mo (110) surface, the O atom initially placed at the Bridge site of the Mo-Re (110) surface will move to the Hollow site after the structure relaxation. In addition, the O atom can be stabilized at the Top site with a much higher adsorption energy of −2.82 eV. As with Na adsorption, the Re atom in the surface can also strengthen the attractive interaction between the adsorbed O atom and the Mo-based substrate.

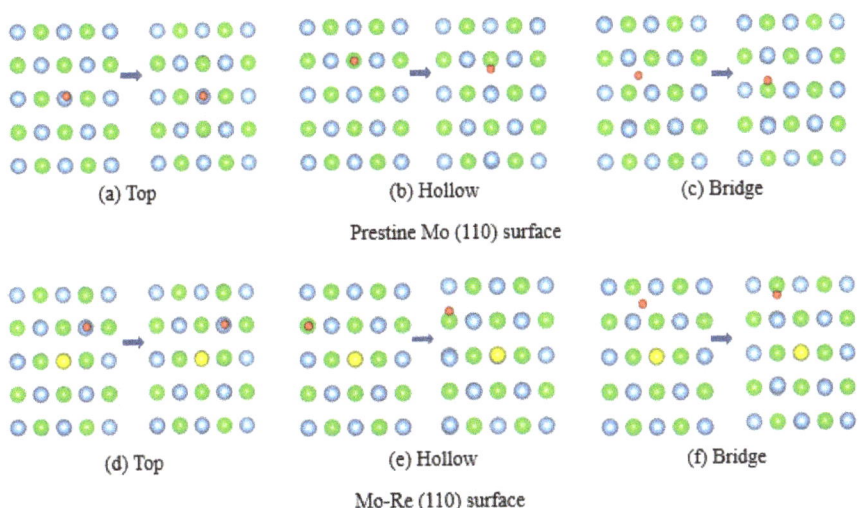

Figure 5. Initial and final configurations of O adsorption on Mo (110) and Mo-Re (110) surfaces. Scheme 110. surface. Snaps (**a**–**c**) show O atoms initially placed at Top site, Hollow site and Bridge site of Mo (110) surface, while snaps (**d**–**f**) show O atoms initially placed at Top site, Hollow site and Bridge site of Mo-Re (110) surface. Blue and green spheres represent Mo atoms in the first layer and the second layer. Yellow and red spheres represent Re atoms and O atoms, respectively.

3.3. Impact of O on Na Adsorption and Diffusion

O is the key impurity in liquid metal for the high-temperature heat pipe. The impact of pre-adsorbed O on Na adsorption behavior is also investigated. The configurations of pre-adsorbed O atom are adopted from Figure 6b,e. The adsorption energy for the Na on surface with a pre-adsorbed O is defined as

$$E_{ads} = E_{(Mo-O/Mo-Re-O)-Na} - E_{(Mo-O/Mo-Re-O)slab} - E_{Na} \qquad (4)$$

where $E_{(Mo-O/Mo-Re-O)-Na}$ is the total energy of the adsorbate–substrate system, $E_{(Mo-O/Mo-Re-O)slab}$ is the total energy of Mo or Mo-Re surface with a pre-adsorbed O atom, and E_{Na} is the energy of an Na atom in the BCC structure. In Equation (4), the subscript Mo-O represents the Mo (110) surface with a pre-adsorbed O atom and Mo-Re-O represent the Mo-Re (110) surface with a pre-adsorbed O atom.

All un-equivalent sites are considered and only the configurations with the lowest energies are shown in Figure 6. Adsorption energies and geometric parameters are given in Table 4. It is found that Na atoms initially placed at Top and Bridge sites move to Hollow sites after the optimization. The former one occupied the Hollow site which is 6.02 Å away from the pre-adsorbed O atom, and latter one occupied the Hollow site which is only 2.35 Å away from the O atom. However, these two final configurations lead to the same adsorption energy of 0.52 eV. For the Na atom initially placed at the Hollow site, the adsorption energy is −0.53 eV with $d_{Na-O} = 2.36$ Å. It can be inferred that the O atom does not affect the adsorption behavior of the Na atom.

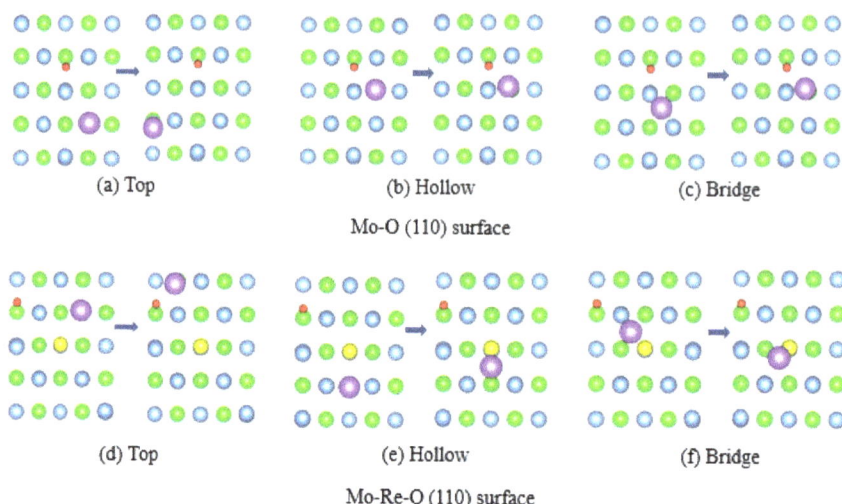

Figure 6. The initial and final configurations of Na atom adsorption on Mo-O (110) and Mo-Re-O (110) surface. Snaps (a–c) show Na atoms initially placed at Top site, Hollow site and Bridge site of Mo-O (110) surface, while snaps (d–f) show Na atoms initially placed at Top site, Hollow site and Bridge site of Mo-Re-O (110) surface. Blue and green spheres represent Mo atoms in the first layer and the second layer. Yellow, purple and red spheres represent Re atoms, Na atoms and O atoms, respectively.

Table 4. Na adsorption energy on different surface models.

Surface	Initial Site	d_{Na-O} (Å)	d_{Na-sur} (Å)	E_{ads} (eV)	Final Stie
	Top	6.02	2.78	−0.52	Hollow
Mo-O (110)	Hollow	2.36	2.68	−0.53	Hollow
	Bridge	2.35	2.83	−0.52	Hollow
	Top	2.37	2.76	−0.54	Hollow
Mo-Re-O (110)	Hollow	4.77	2.71	−0.61	Hollow
	Bridge	3.18	2.77	−0.57	Bridge

For the Mo-Re (110) surface with a pre-adsorbed O, Na initially placed at the top site will spontaneously move to a Hollow site which is close to the O atom (d_{Na-O} = 2.37 Å) as shown in Figure 6d. However, it should be noticed that the Na atom at the Hollow site which is closer to a Re atom has lowest adsorption energy of −0.61 eV in Figure 6e. The adsorption energy of Na at the Bridge site is −0.57 eV, which is also closer to the Re atom in Figure 6f and has a lower adsorption energy than the Na atom shown in Figure 6d. For the Na adsorption on the clean Mo-Re (110) surface, the adsorption energies of the Hollow site and Bridge site are −0.56 eV and −0.60 eV. As with the Mo (110) surface, pre-adsorbed O on the Mo-Re (110) surface cannot affect the Na adsorption behavior significantly.

Figure 7 show the energy barrier of Na migration from one most stable site to its first-nearest most stable site is also calculated in this work using CI-NEB method. Our theoretical results show that the diffusion barrier of Na on Mo (110) surface is 0.037 eV, while it is 0.063 eV on the Mo-Re (110) surface. It can be inferred that the Re atom can slower down the Na diffusion kinetics on the Mo surface. The impact of pre-adsorbed O on the Na diffusion is also investigated. For the Mo (110) surface, the pre-adsorbed O atom can increase the diffusion barrier to 0.087; for the Mo-Re (110) surface, the pre-adsorbed O can significantly increase the Na diffusion barrier to 0.221 eV. Therefore, it can be inferred that both the existence of O impurity and Re alloy atoms can block the Na diffusion on the Mo surface.

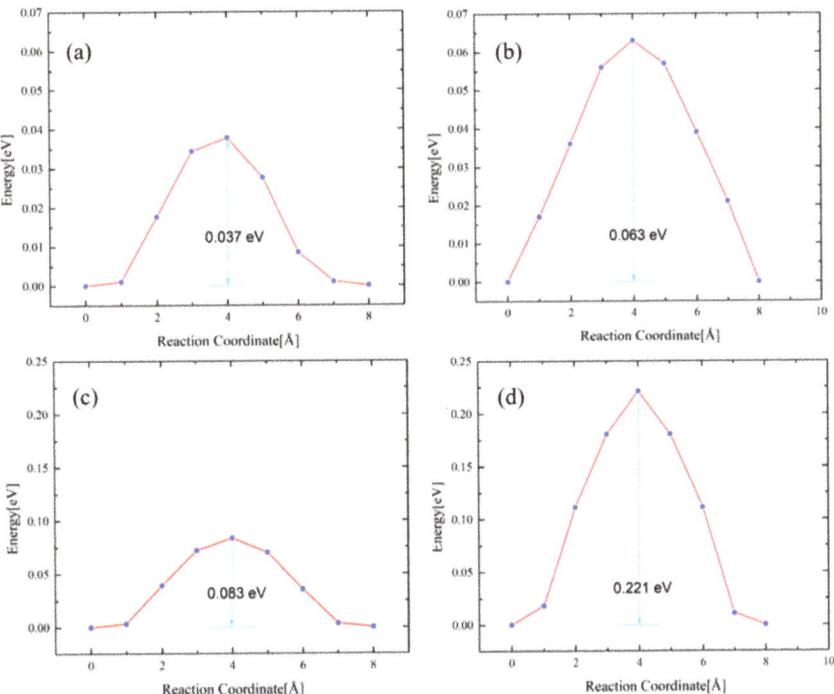

Figure 7. Calculated diffusion energy profiles for a Na atom diffusion on different surfaces: (**a**) Na diffusion on pure Mo (110) surface; (**b**) Na diffusion on Mo-Re (110) surface; (**c**) Na diffusion on Mo (110) surface with an adsorbed O atom; (**d**) Na diffusion on Mo-Re (110) surface with an adsorbed O atom.

3.4. Impact of O/Na atom Adsorption on the Vacancy Formation of Mo (110) and Mo-Re (110) Surfaces

We propose that Mo dissolution into liquid Na is a potential scenario of Mo corrosion. In this study, the Mo vacancy formation energy is also calculated to evaluate the effect of O and Re on the corrosion resistance of Mo surface. The vacancy formation energy is defined using the following Equation (5):

$$E_{vac}^{f} = E_{vac-surface} + E_{Mo} - E_{surface} \quad (5)$$

where $E_{surface}$ stands for the energy of surfaces without vacancies, E_{Mo} is the energy of a Mo atom in its BCC structure and $E_{vac-surface}$ is the surface models with a Mo vacancy in the upmost layer.

Table 5 shows that the vacancy formation energies are 1.45 eV and 1.32 eV for the clean Mo (110) surface and the clean Mo-Re (110) surface, respectively. Therefore, introducing Re atom into the Mo (110) surface can enhance forming surface vacancies. The surface vacation formation energies of Mo (110) or Mo-Re (110) surface with an Na or O atom is also calculated. A single Na atom or O atom can reduce the surface vacancy formation energy by 0.4~0.5 eV. It can be inferred that a low coverage of $\Theta = \frac{1}{8}$ ML of Na or O can destabilize the Mo (110) and Mo-Re (110) surface. The effect of Na/O synergetic effect on vacancy formation is also studied. The present theoretical results show that the vacancy formation energy is significantly decreased when a Na atom and a O atom are co-adsorbed on the Mo (110) surface. Therefore, it can be inferred that impurity can facilitate the dissolution of Mo atoms and lead to corrosion. For the Na/O co-adsorption condition, the existence of Re atom can increase the surface vacancy formation energy to 1.06 eV. Hence, Re can prevent the Mo alloy from corrosion in liquid Na with O impurities.

Table 5. Mo vacancy formation energy on different surface models.

Adsorbate	Surface	Vacancy Formation Energy (eV)
Clean	Mo (110)	1.45
	Mo-Re (110)	1.32
A single O atom	Mo (110)	0.97
	Mo-Re (110)	0.86
A single Na atom	Mo (110)	0.93
	Mo-Re (110)	0.90
Na and O co-adsorption	Mo (110)	0.47
	Mo-Re (110)	1.06

4. Conclusions

In this study, the adsorption and diffusion behaviors of Na and O atoms on BCC-Mo (110) surface are investigated by a first-principles approach. It is found that the Hollow site is the most energetically preferred adsorption site for a single Na or O atom. The existence of a Re alloy atom in the first atomic layer of the surface can strengthen the attractive interaction between the adsorbate and the substrate. The diffusion barrier of the Na atom on the Mo (110) surface is only 0.037 eV, and the pre-adsorbed O atom and Re atom can significantly impede Na diffusion. The surface vacancy formation energy is calculated for evaluating the stability of the Mo (110) surface. It is found that the Na or O atom can decrease the formation energy of the surface vacancy. It is worth noting that the Na/O co-adsorption can significantly reduce the vacancy formation energy to 0.47 eV for the Mo (110) surface, which indicates that the dissolution of surface atoms is a potential mechanism for the Mo in the Na liquid with O impurity. However, Re as alloy element can increase the resistance to the dissolution induced by Na/O co-adsorption.

Author Contributions: Q.Z. performs simulation and data analyzation, and writes the original draft; Z.L. concept this work; W.L., M.M. and H.D. refine the manuscript. All authors have read and agreed to the published version of the manuscript.

Funding: This study is financially supported by the President Foundation of China Academy of Engineering Physics (YZJJLX2018002).

Institutional Review Board Statement: Not applicable.

Informed Consent Statement: Not applicable.

Data Availability Statement: Data sharing is not applicable to this article.

Conflicts of Interest: The authors declare no conflict of interest.

References

1. Zhang, H.; Zhuang, J. Research, development and industrial application of heat pipe technology in China. *Appl. Therm. Eng.* **2003**, *23*, 1067–1083. [CrossRef]
2. Ai, B.; Chen, S.; Yu, J.; Lu, Q.; Han, H.; Hu, L. Fabrication of lithium/C-103 alloy heat pipes for sharp leading edge cooling. *Heat Mass Transf.* **2017**, *54*, 1359–1366. [CrossRef]
3. Wang, C.; Guo, Z.; Zhang, D.; Qiu, S.; Tian, W.; Wu, Y.; Su, G. Transient behavior of the sodium–potassium alloy heat pipe in passive residual heat removal system of molten salt reactor. *Prog. Nucl. Energy* **2013**, *68*, 142–152. [CrossRef]
4. Faghri, A. Review and Advances in Heat Pipe Science and Technology. *J. Heat Transf.* **2012**, *134*, 123001. [CrossRef]
5. Reay, D.; McGlen, R.; Kew, P. *Heat Pipes: Theory, Design and Applications*, 6th ed.; Elsevier: Amsterdam, The Netherlands, 2014.
6. Jouhara, H.; Chauhan, A.; Nannou, T.; Almahmoud, S.; Delpech, B.; Wrobel, L. Heat pipe based systems—Advances and applications. *Energy* **2017**, *128*, 729–754. [CrossRef]

7. Qu, W. Progress Works of High and Super High Temperature Heat Pipes. In *Developments in Heat Transfer*; Bernardes, M.A.D.S., Ed.; InTech: London, UK, 2011; pp. 503–522. ISBN 978-953-307-569-3.
8. El-Genk, M.; Tournier, J.-M. Challenges and fundamentals of modeling heat pipes' startup from a frozen state. *AIP Conf. Proc.* **2002**, *608*, 127–138.
9. Lundberg, L.B. Refractory Metals in Space Nuclear Power. *JOM* **1985**, *37*, 44–47. [CrossRef]
10. Lundberg, L. An evaluation of molybdenum and its alloys. In Proceedings of the 16th Thermophysics Conference, American Institute of Aeronautics and Astronautics (AIAA), Palo Alto, CA, USA, 23–25 June 1981.
11. King, J.C.; El-Genk, M.S. Review of Refractory Materials for Alkali Metal Thermal-to-Electric Conversion Cells. *J. Propuls. Power* **2001**, *17*, 547–556. [CrossRef]
12. El-Genk, M.S.; Tournier, J.-M. A review of refractory metal alloys and mechanically alloyed-oxide dispersion strengthened steels for space nuclear power systems. *J. Nucl. Mater.* **2005**, *340*, 93–112. [CrossRef]
13. Hampel, C. Refractory Metals. Tantalum, Niobium, Molybdenum, Rhenium, and Tungsten. *Ind. Eng. Chem.* **1961**, *53*, 90–96. [CrossRef]
14. Shmelev, A.N.; Kozhahmet, B.K. Use of molybdenum as a structural material of fuel elements for improving the safety of nuclear reactors. *J. Phys. Conf. Ser.* **2017**, *781*, 012022. [CrossRef]
15. Gilbert, M.; Packer, L.; Stainer, T. Experimental validation of inventory simulations on molybdenum and its isotopes for fusion applications. *Nucl. Fusion* **2020**, *60*, 106022. [CrossRef]
16. Lundberg, L. *Critical Evaluation of Molybdenum and Its Alloys for Use in Space Reactor Core Heat Pipes*; No. LA-8685-MS; Los Alamos Scientific Laboratory: Los Alamos, NM, USA, 1981. [CrossRef]
17. Yu, X.; Kumar, K. Uniaxial, load-controlled cyclic deformation of recrystallized molybdenum sheet. *Mater. Sci. Eng. A* **2012**, *540*, 187–197. [CrossRef]
18. Jörg, T.; Music, D.; Hauser, F.; Cordill, M.J.; Franz, R.; Köstenbauer, H.; Winkler, J.; Schneider, J.; Mitterer, C. Deformation behavior of Re alloyed Mo thin films on flexible substrates: In situ fragmentation analysis supported by first-principles calculations. *Sci. Rep.* **2017**, *7*, 7374. [CrossRef]
19. Agnew, S.R.; Leonhardt, T. The low-temperature mechanical behavior of molybdenum-rhenium. *JOM* **2003**, *55*, 25–29. [CrossRef]
20. DiStefano, J.R. Review of alkali metal and refractory alloy compatibility for Rankine cycle applications. *J. Mater. Eng.* **1989**, *11*, 215–225. [CrossRef]
21. Tu, S.-T.; Zhang, H.; Zhou, W.W. Corrosion failures of high temperature heat pipes. *Eng. Fail. Anal.* **1999**, *6*, 363–370. [CrossRef]
22. Kresse, G.; Furthmüller, J. Efficiency of ab-initio total energy calculations for metals and semiconductors using a plane-wave basis set. *Comput. Mater. Sci.* **1996**, *6*, 15–50. [CrossRef]
23. Kresse, G.; Furthmüller, J. Efficient Iterative Schemes for Ab Initio Total-Energy Calculations Using a Plane-wave Basis Set. *Phys. Rev. B* **1996**, *54*, 11169–11186. [CrossRef]
24. Car, R.; Parrinello, M. Unified Approach for Molecular Dynamics and Density-Functional Theory. *Phys. Rev. Lett.* **1985**, *55*, 2471–2474. [CrossRef]
25. Jones, R.O.; Gunnarsson, O. The density functional formalism, its applications and prospects. *Rev. Mod. Phys.* **1989**, *61*, 689–746. [CrossRef]
26. Payne, M.C.; Teter, M.P.; Allan, D.C.; Arias, T.A.; Joannopoulos, A.J. Iterative minimization techniques for ab initio total-energy calculations: Molecular dynamics and conjugate gradients. *J. Rev. Mod. Phys.* **1992**, *64*, 1045. [CrossRef]
27. Blöchl, P.E. Projector augmented-wave method. *Phys. Rev. B* **1994**, *50*, 17953–17979. [CrossRef] [PubMed]
28. Kresse, G.; Joubert, D. From ultrasoft pseudopotentials to the projector augmented-wave method. *Phys. Rev. B* **1999**, *59*, 1758. [CrossRef]
29. Perdew, J.P.; Burke, K.; Ernzerhof, M. Generalized Gradient Approximation Made Simple. *Phys. Rev. Lett.* **1996**, *77*, 3865. [CrossRef]
30. Monkhorst, H.J.; Pack, J.D. Special points for Brillouin-zone integrations. *Phys. Rev. B* **1976**, *13*, 5188–5192. [CrossRef]
31. Jain, A.; Ong, S.P.; Hautier, G.; Chen, W.; Richards, W.D.; Dacek, S.; Cholia, S.; Gunter, D.; Skinner, D.; Ceder, G.; et al. Commentary: The Materials Project: A materials genome approach to accelerating materials innovation. *APL Mater.* **2013**, *1*, 011002. [CrossRef]
32. Okamoto, H. Supplemental Literature Review of Binary Phase Diagrams: Ag-Yb, Al-Co, Al-I, Co-Cr, Cs-Te, In-Sr, Mg-Tl, Mn-Pd, Mo-O, Mo-Re, Ni-Os, and V-Zr. *J. Phase Equilibria Diffus.* **2016**, *37*, 726–737. [CrossRef]
33. Henkelman, G.; Uberuaga, B.; Jónsson, H. A climbing image nudged elastic band method for finding saddle points and minimum energy paths. *J. Chem. Phys.* **2000**, *113*, 9901–9904. [CrossRef]
34. American Society for Metals and ASM Handbook Committee. *Properties and Selection: Nonferrous Alloys and Pure Metals*, 9th ed.; American Society for Metals: Russell Township, OH, USA, 1979.
35. Guo, F.; Wang, J.; Du, Y.; Wang, J.; Shang, S.-L.; Li, Y.; Chen, L. First-principles study of adsorption and diffusion of oxygen on surfaces of TiN, ZrN and HfN. *Appl. Surf. Sci.* **2018**, *452*, 457–462. [CrossRef]
36. Graciani, J.; Sanz, J.F.; Asaki, T.; Nakamura, K.; Rodriguez, J.A. Interaction of oxygen with TiN(001):N↔O exchange and oxidation process. *J. Chem. Phys.* **2007**, *126*, 244713. [CrossRef] [PubMed]
37. Osei-Agyemang, E.; Balasubramanian, G. Surface oxidation mechanism of a refractory high-entropy alloy. *Npj Mater. Degrad.* **2019**, *3*, 20. [CrossRef]

Article

Atomic Simulations of U-Mo under Irradiation: A New Angular Dependent Potential

Wenhong Ouyang, Wensheng Lai, Jiahao Li, Jianbo Liu * and Baixin Liu

The Key Laboratory of Advanced Materials (MOE), School of Materials Science and Engineering, Tsinghua University, Beijing 100084, China; oywh18@mails.tsinghua.edu.cn (W.O.); wslai@mail.tsinghua.edu.cn (W.L.); lijiahao@mail.tsinghua.edu.cn (J.L.); dmslbx@mail.tsinghua.edu.cn (B.L.)
* Correspondence: jbliu@mail.tsinghua.edu.cn

Abstract: Uranium-Molybdenum alloy has been a promising option in the production of metallic nuclear fuels, where the introduction of Molybdenum enhances mechanical properties, corrosion resistance, and dimensional stability of fuel components. Meanwhile, few potential options for molecular dynamics simulations of U and its alloys have been reported due to the difficulty in the description of the directional effects within atomic interactions, mainly induced by itinerant f-electron behaviors. In the present study, a new angular dependent potential formalism proposed by the author's group has been further applied to the description of the U-Mo systems, which has achieved a moderately well reproduction of macroscopic properties such as lattice constants and elastic constants of reference phases. Moreover, the potential has been further improved to more accurately describe the threshold displacement energy surface at intermediate and short atomic distances. Simulations of primary radiation damage in solid solutions of the U-Mo system have also been carried out and an uplift in the residual defect population has been observed when the Mo content decreases to around 5 wt.%, which corroborates the negative role of local Mo depletion in mitigation of irradiation damage and consequent swelling behavior.

Keywords: uranium; molybdenum; molecular dynamics; interatomic potential

1. Introduction

Pure uranium (U) is limited to a considerable extent in the production of fuel elements due to its questionable properties. α-U, which is stable at low temperatures, has pronounced anisotropy and comparatively low strength characteristics at elevated temperatures. Meanwhile, the existence of high-temperature body-centered cubic (bcc) γ-U and intermediate tetragonal β-U indicates harmful allotropic transformations during cyclic heat treatment. Moreover, U shows high chemical activity in corrosion behaviors [1,2]. A common solution of problems above is the utilization of uranium alloys, which retains the isotropic structure of γ-U at room temperature. So far, molybdenum (Mo) and zirconium (Zr), along with a few other elements appear to be promising options in U alloys for nuclear fuel. In particular, alloying U with Mo remarkably improves its mechanical properties, corrosion resistance, and dimensional stability, along with providing other benefits such as high thermal conductivity and low thermal expansion [3,4]. Enhanced mechanical properties enable U-Mo alloys to be utilized for the production of cores of fuel elements of arbitrary configurations. High corrosion resistance in water of high parameters and reliable dimensional stability under irradiation are deciding factors governing the choice of U-Mo alloys as fuel materials. It is also deemed expedient in all cases for U-Mo alloys to be included in the construction of fast-neutron reactors [1].

However, the introduction of alloy elements also brings about some peculiar physical phenomena whose fundamental mechanism remains to be understood. During the cooling of U-Mo alloys, several metastable phases have been observed in a sequence of bcc γ → bcc with doubled lattice constant $γ^S$ → body-centered tetragonal (bct) $γ^0$ → monoclinic α″ → orthorhombic α′ [5,6]. The addition of Mo was suggested to stiffen the U lattice against shear, thus hindering and also complicating the transition progress. However, during service life, Mo has been observed to be depleted from grain boundaries (GBs) among a wide range of compositional banding [7], which could lead to an onset of phase decomposition. In particular, the elemental redistribution near GBs could also be implicated in the generation of irradiation-induced recrystallization (IIR) [8], where fuel grains are subdivided into nano-sized grains from the GBs during service life. Meanwhile, IIR is suggested as an important culprit behind accelerated swelling behavior of nuclear fuel alloys [9], which enhances the reach of GBs into the fuel grains, destroying low swelling intra-granular fission gas bubbles (FGBs) and producing high swelling inter-granular FGBs.

To shed light on the underlying mechanism of microstructure evolution under irradiation at U-Mo fuel operation temperatures, constructing corresponding potential for molecular dynamics simulation has been under research in recent years. An embedded atom method (EAM) potential for U-Mo-Xe has been developed by the force matching method [10], through which the properties of U2Mo and α-U-Mo could be reproduced well, even without taking them into account in the fitting process. However, traces of overfitting were observed, suggesting the limitation of EAM for the description of a U-Mo system. By the use of the interatomic potential proposed by Smirnova et al. [10], attempts to reveal cascade effects on residual defects in U-Mo alloys have also been reported [11]. Also based on the U-Mo-Xe EAM potential, Hu et al. [12] have investigated the relationship between the pressure, equilibrium Xe concentration, and radius of Xe bubbles in U-10 wt% Mo by molecular dynamics (MD) simulations. Utilizing a formalism improved on the basis of EAM, namely Angular-Dependent Potential (ADP), Smirnova et al. have qualitatively reproduced the properties of cubic and tetragonal phases of γ-U-Mo alloys [13] and improvements were also made in reproduction of the density, coefficient of thermal expansion, and diffusion behavior [14]. It was also suggested that a successful capture of atomic properties in γ-U and γ-U-Mo systems requires utilization of a potential form, with its level of complexity no lower than the ADP. Moreover, Starikov et al. [15] have also constructed an ADP with the same form as that developed by Smirnova, which paid special attention to the description of metastable phases of U-Mo solid solution in the fitting process.

For the present work, we first constructed a U-Mo interatomic potential using the ADP formalism proposed by the author's group [16]. The ADP formulas were further modified such that it can more accurately reproduce the threshold displacement energy surface as well as many-body repulsion at intermediate and short interatomic distances. We then applied the obtained potential in MD simulations to study the U-Mo system under irradiation.

2. Materials and Methods

Based on the formalism of EAM, several modified versions including the MEAM and ADP potentials were developed by the addition of angular dependent terms to increase the description accuracy. Similarly, for the present study, the new angular-dependent interatomic potential calculated atomic energy contribution E_i by the following formula:

$$E_i = \frac{1}{2}\sum_{j \neq i}\phi_{ij}(r_{ij}) + F(\overline{\rho}_i) - \vartheta_i \tag{1}$$

where

$$\phi_{ij}(r_{ij}) = \sum_{i=0}^{3} a_i(r_c - r_{ij})^{3+i},$$
$$F(\overline{\rho}_i) = -\sqrt{\overline{\rho}_i}, \overline{\rho}_i = \sum_{j \neq i}\rho_{ij}(r_{ij}),$$
$$\rho_{ij}(r_{ij}) = a^2(r_c - r_{ij})^4 \exp\left[-\beta\left(\frac{r_{ij}}{r_0} - 1\right)^2\right], \vartheta_i = \vartheta_i^u + \vartheta_i^v + \vartheta_i^w. \tag{2}$$

It could be indicated by the first two terms in Equation (1) that the main part of the atomic energy maintains the form of the widely used EAM potential. To be exact, ϕ_{ij} and

$F(\bar{\rho}_i)$ describe the atomic interactions of the electrostatic repulsion between ion cores and the attraction induced by embedding atoms in the electron atmosphere provided by their neighbor atoms, respectively. In calculation of E_i of the i-th atom, summation is over its j-th neighbor atom within the cutoff distance r_c, which was set to 4.7 Å in the present study. In particular, r_0 and r_{ij} denotes the first neighbor distance in a reference lattice at equilibrium and the distance between the i-th center atom and its j-th neighbor atom, respectively. Due to the inherent sphere symmetry in the potential form, EAM poorly describes some peculiar atomic behaviors in systems of U alloys, such as the stability of the alpha phase with low symmetry at room temperature. In the present work, the description of the atomic energy was modified with the addition of a new angular-dependent term ϑ_i, which has three components given as the following:

$$\vartheta_i^u = \sum_{\alpha} \left(\sum_{j \neq i} \psi_j^u(r_{ij}) \frac{r_{ij}^{\alpha}}{r_{ij}} \right)^2,$$

$$\vartheta_i^v = \sum_{\alpha,\beta} \left(\sum_{j \neq i} \psi_j^v(r_{ij}) \frac{r_{ij}^{\alpha} r_{ij}^{\beta}}{r_{ij} r_{ij}} \right)^2 - \frac{1}{3}\left(\sum_{j \neq i} \psi_j^v(r_{ij}) \right)^2, \quad (3)$$

$$\vartheta_i^w = \sum_{\alpha,\beta,\gamma} \left(\sum_{j \neq i} \psi_j^w(r_{ij}) \frac{r_{ij}^{\alpha} r_{ij}^{\beta} r_{ij}^{\gamma}}{r_{ij} r_{ij} r_{ij}} \right)^2 - \frac{3}{5}\sum_{\alpha}\left(\sum_{j \neq i} \psi_j^w(r_{ij}) \frac{r_{ij}^{\alpha}}{r_{ij}} \right)^2$$

where

$$\psi_j^t(r) = c_t \exp\left(-\frac{(r-d_t)^2}{2\lambda_t^2} \right), t = u, v, w. \quad (4)$$

In Equation (3), superscripts $\alpha, \beta, \gamma = 1, 2, 3$ refer to the Cartesian components of position vectors. ϑ_i^u, ϑ_i^v, and ϑ_i^w denote the angular-dependent components of the first, second, and third order, with ψ_j^t as basis function in the form of the normal distribution function. In the present study, it was assumed that all the three angular-dependent components are required in the fitting process for the description of U and the first two ones for Mo.

A total of 15 parameters were adjusted in the fitting process, which include $a_i, \alpha, \beta, c_t, d_t, \lambda_t$ listed in Table 1. In particular, the angular-dependent parameters of the third order for Mo were kept to zero. During the fitting, an error function was optimized. Macroscopic properties including the cohesive energy, lattice parameters, and elastic constants of certain lattices were calculated with a set of fitting parameters and compared with reference data from the literature or calculations by the First Principle (FP) method. The framework of the fitting procedure was implemented in Python, where symbolic computation and optimization are supported by Sympy and Scipy packages. In particular, in order to improve the description of point defects, several lattices with certain point defects were included in the reference data and allowed to relax during the calculations of the formation energies.

In order to capture atomic behaviors within equilibrium distances, ϕ_{ij} and $\rho_{ij}(r_{ij})$ within the inner cutoff were further modified. Interpolations are implemented for $\rho_{ij}(r_{ij})$ to approach a constant and for ϕ_{ij} to connect smoothly with the well-known Ziegler-Biersack-Littmark (ZBL) potential [17]. In particular, special attention was paid to the intermediate repulsive range in development of the potential for further employment in cascade simulations. Traditionally speaking, a smooth interpolation is used between the pairwise part of many-body potentials in the near-equilibrium range and the ZBL potential in the short range. Moreover, in 2016, Stoller et al. [18] pointed out that no accepted standard method had been developed in the interpolation process and that how the force fields are linked can sensibly influence the results of cascade simulation. Similar to the modifications implemented by Stoller, the present study also corrected the potential with calculation results from density functional theory (DFT) as a benchmark. The DFT calculations were performed using the Vienna ab initio simulation package (VASP) [19] with projector augmented wave pseudo-potentials. A kinetic energy cutoff of 400 eV and a $4 \times 4 \times 4$ grid with the Monkhorst–Pack method were employed. The ADP calculations were performed using the Large-scale Atomic/Molecular Massively Parallel Simulator (LAMMPS) [20]. A supercell containing $3 \times 3 \times 3$ conventional bcc cells under equilibrium

($a_U = 3.48$ Å) was chosen for calculations of energetic benchmark and then an atom in the supercell was translated in several equidistant steps toward its neighbors along the direction of <110> and <100>.

Table 1. Fitted parameters of ADP for U–Mo system.

Parameter	U–U	Mo–Mo	U–Mo
a_0 (eV·Å^{-3})	−1.41881	1.75876	6.095612
a_1 (eV·Å^{-4})	2.499597	−3.74227	−11.667
a_2 (eV·Å^{-5})	−1.40956	2.276131	7.291495
a_3 (eV·Å^{-6})	0.279328	−0.42748	−1.48751
α (eV·Å^{-2})	0.86737	0.255577	-
β	−0.88394	2.017747	-
c_u (eV$^{1/2}$)	0.083778	0.369529	−0.2
d_u (Å)	2.825854	2.590854	3.173449
λ_u (Å)	0.151763	0.62249	0.312741
c_v (eV$^{1/2}$)	0.015177	0.225399	0.2
d_v (Å)	3.153919	2.791784	3.208256
λ_v (Å)	0.149995	0.186653	0.409074
c_w (eV$^{1/2}$)	0.133803	-	−0.2
d_w (Å)	3.402176	-	2.952746
λ_w (Å)	0.359213	-	0.167571
r_c (Å)	4.7	4.7	4.7
r_0 (Å)	2.7408	2.7387	-

For simplicity in description of energetic variance, the present study used ΔE_t as the following:

$$\Delta E_t = E_{\text{translated}} - E_{\text{perfect}} \quad (5)$$

where $E_{\text{translated}}$ is the energy of the configuration with atomic translation and E_{perfect} is the energy of the original configuration with perfect symmetry. The energetics of the supercell as a function of the ratio between the distance of the nearing dimers and the lattice constant are shown in Figure 1. It could be seen that a reasonable reproduction of energetics variance was obtained in the intermediate range, especially for U-U and U-Mo dimers. Moreover, it should be noted that in cases of dimers with different elemental components, the choice of which atom is moved toward the other produces similar but actually different energetics variances. The difference depends mainly on the distinct interactions between the translated atom and its atomic environment. Good reproduction was achieved in the present study, indicating reasonable accuracy of the refitted potential in the intermediate range. Meanwhile, sensible deviations could be seen at the outer end of the intermediate range, especially for Mo-Mo dimers, which are induced by the difficulty of resolving the energetic uplift at the outer end of the intermediate range while leaving the description by the ADP of near-equilibrium atomic interactions intact. Nonetheless, even under the deviations, the general curve shapes of energetics variance were reserved to a large extent, indicating a good description of dynamic properties during atomic interactions in the intermediate range.

The newly developed potential was then applied in MD simulations of displacement cascades in U-Mo alloys in the present study. Simulation boxes containing $60 \times 60 \times 60$ bcc unit cells were set up under periodic boundary conditions (PBC) with up to 648,000 atoms included. Alloying elements were randomly distributed in initial lattices before simulations. A total of four proportions of alloying elements, that is, 0 wt.%, 2.5 wt.%, 5 wt.%, 7.5 wt.%, and 10 wt.%, were chosen to evaluate the effect of the content of the alloying element on displacement cascades. Given the high symmetry of the bcc lattice, a total of 15 evenly distributed directions were sampled in the fundamental orientation zone, which is shown in Figure 2. The initial kinetic energies of the primary knock-on atoms (PKA) was set to 5 keV for all the simulations. For each set of initial conditions, simulations were carried out up to eight times to reduce statistical error in the following analysis.

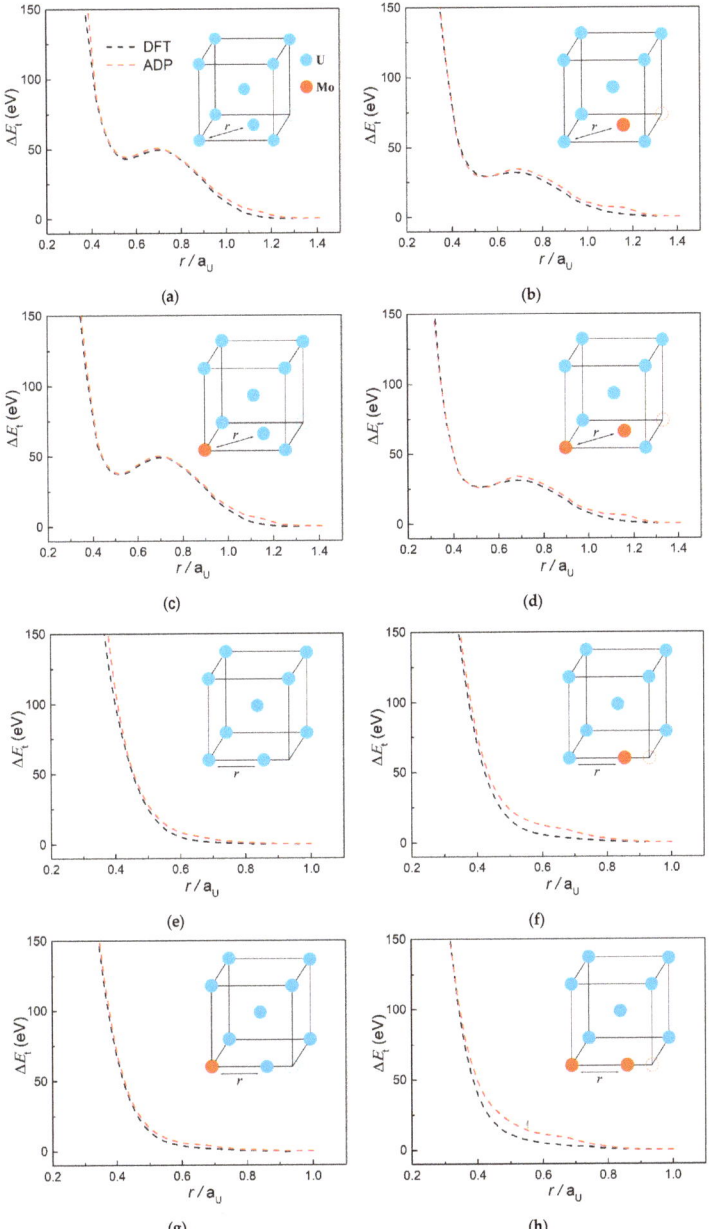

Figure 1. The energetics of the supercell as a function of the ratio between the distance of the nearing dimers and the lattice constant of bcc U. In particular, nearing atom pairs include (**a**) U–U in <110>, (**b**) U–Mo in <110>, (**c**) Mo–U in <110>, (**d**) Mo–Mo in <110>, (**e**) U–U in <100>, (**f**) U–Mo in <100>, (**g**) Mo–U in <100> and (**h**) Mo–Mo in <100>. The black and red dash lines represent calculation results from DFT and MD using ADP. In the configuration diagrams, blue and orange circles represent U and Mo atoms, respectively, and the original positions of displaced atoms are indicated by dotted circles.

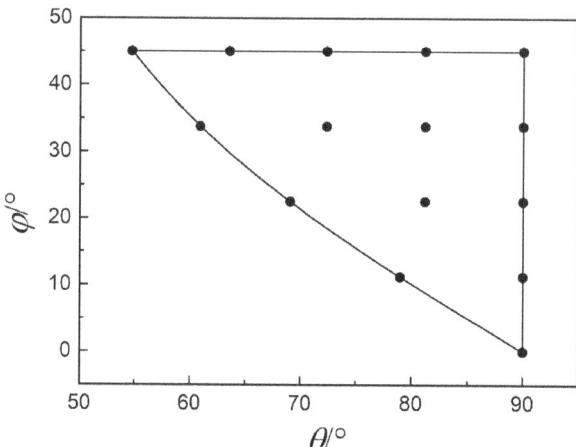

Figure 2. Angular distribution of launch directions sampled in the fundamental orientation zone of bcc lattice.

Prior to each set of cascade simulations, the simulation block was equilibrated under 600 K and 0 GPa in an isothermal-isobaric (NPT) ensemble until the temperature and pressure reached a stable level. The initial structures for cascade simulations were chosen randomly from those under equilibrium in a series of time steps. During the cascade simulations, the simulation block was divided into three layers. In the outmost layer, the atoms within 2 Å from the boundary were maintained as static all along the simulation to avoid excessive overall displacement of the simulation block. Atoms located at a distance of 2 to 8 Å from the boundary were partitioned into the second layer. In this layer, a thermostat was applied by rescaling the velocities of the atoms to maintain a stable level of temperature and to absorb the excess kinetic energy from the cascade. The rest of atoms in the core region were simulated with the constant NVE ensemble and a variable time step in a range of 0.01 to 1 fs was used in all simulations, which depends on the maximum of atom displacements between consecutive steps.

To determine and record defect sites during simulations, some native commands in LAMMPS were invoked, which construct a voronoi tessellation dividing each atom into an exclusive cell at the start of each simulation. In particular, at equal intervals of timesteps, voronoi cells with occupancy not equal to 1 were recorded with their position coordinates, while those with occupancy equal to 0 and larger than 1 were marked as vacancy and interstitial sites, respectively. In cluster analysis, adjacent defect sites within a cutoff radius of 4.1 Å are classified into the same cluster group and the size of a cluster is defined as the total count of defect sites within the cluster. By using other commands in LAMMPS, atom displacements and element types were also recorded. The program OVITO [21] was utilized for data extraction and statistical analysis.

3. Results

3.1. Fitting Results

In the fitting procedure, reference data were gathered from experimental results in the literature and FP calculations as supplements, especially for phases not observed in experiments. Fitted parameters are shown in Table 1. The functions defining the newly developed ADP are plotted in Figure 3.

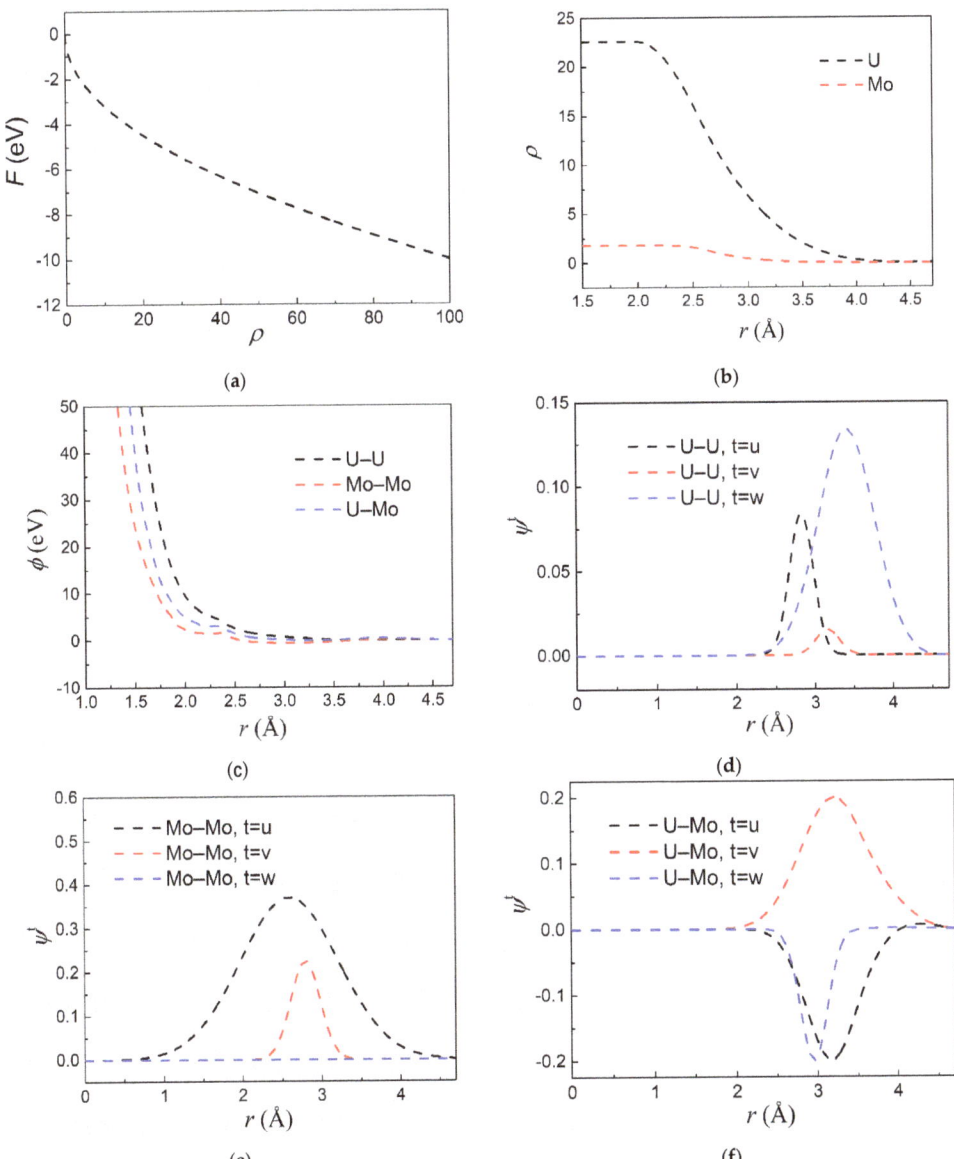

Figure 3. Potential functions of the newly developed ADP. Embedding function F, electron density function ρ and pair potential ϕ are shown in (**a**–**c**), respectively; angular-dependent terms of ψ^u, ψ^v and ψ^w are grouped together and plotted for elemental pairs of (**d**) U–U, (**e**) Mo–Mo and (**f**) U–Mo.

Fitting results of pure phases of U and Mo are shown in Tables 2 and 3. It could be seen that the lattice parameters were reproduced moderately well and the hierarchy of relative stability of allotropes was correctly reflected in terms of the cohesive energies. During fitting, specific attention was paid to the reproduction of elastic constants and the formation energies of defects. The fitting results of cross potentials of U–Mo are given in Table 4. During the fitting of the U–Mo cross potential, two configurations were added to the database, since only the tetragonal phase of U_2Mo has previously been found in experiments.

Table 2. Reproduced values of lattice constants, cohesive energies, and elastic constants of pure uranium.

Structure	Properties	Present Work	MEAM [22]	ADP [15]	Experiment	FP
α-U	E_{coh} (eV/at)	5.46	5.547	4.23	5.550 [23]	-
	a (Å)	2.881	2.721	2.849	2.836 [24]	-
	b (Å)	5.486	6.381	5.841	5.867 [24]	-
	c (Å)	5.221	4.858	4.993	4.935 [24]	-
	y	0.108	0.093	0.103	0.102 [24]	-
	B (GPa)	153	143	147	136 [25]	-
	E_v (eV)	1.17	2.597	-	-	1.95 [26]
γ-U	$\Delta E_{bcc \to ort}$ (eV/at)	0.01	0.15	0.09	-	0.278
	a (Å)	3.479	3.463	3.52	3.47 [27]	3.455
	C_{11} (GPa)	128.2	144.0	183.6	-	103.0
	C_{12} (GPa)	124.1	49.0	92.8	-	142.0
	C_{44} (GPa)	38.2	−36.2	79.9	-	46.0
	E_v (eV)	1.34	-	-	-	1.38
	E_i (eV)	1.08	-	-	-	0.9

Table 3. Reproduced values of lattice constants, cohesive energies, and elastic constants of pure molybdenum.

Structure	Properties	Present Work	FP	Experiment [28–30]
bcc-Mo	E_{coh} (eV/atom)	6.349	6.290	-
	a (Å)	3.171	3.162	3.147
	C_{11} (GPa)	473.9	488.8	465
	C_{12} (GPa)	143.3	146.6	176
	C_{44} (GPa)	67.0	108.3	-
	E_v	2.72	2.723	2.6–3.2
fcc-Mo	$\Delta E_{bcc \to fcc}$ (eV/atom)	0.439	0.327	-
	a (Å)	4.156	4.004	-
hcp-Mo	$\Delta E_{bcc \to hcp}$ (eV/atom)	0.628	0.328	-
	a (Å)	2.842	2.948	-
	c (Å)	4.212	4.263	-

Table 4. Reproduced values of lattice constants, cohesive energies, and elastic constants of several U–Mo binary phases.

Structure	Properties	Present Work	FP
tetra-U_2Mo	E_{coh} (eV/at)	6.044	6.361
	a (Å)	3.304	3.427
	c (Å)	9.921	9.833
	y	0.329	0.328
bcc-$U_{15}Mo$	$\Delta E_{tetra \to bcc}$ (eV/at)	0.391	0.116
	a (Å)	6.861	6.84
	C_{11} (GPa)	159.3	124.4
	C_{12} (GPa)	134.7	140.6
P6-U_2Mo	$\Delta E_{tetra \to P6}$ (eV/at)	48.4	34.9
	a (Å)	−0.119	−0.035
	c (Å)	4.823	4.818

3.2. Cascade Simulations

Typical time evolutions of defects in U-Mo alloy under different Mo contents are given in Figure 4. Similar curves were observed in all cases, in which the counts of defects increase rapidly to a peak value and decrease in a gradually reducing speed. Overall, the addition of Mo increases residual defect populations and Figure 5 shows that with the Mo content, increasing high percentages of residual defects were distributed in large clusters. It could then be inferred that Mo plays a negative role in the recrystallization process of the cascade damage, which in the case of pure U lattice only produces dispersed residual defects in a low population. Considering the inherent impurity of Mo, the recrystallization of bcc lattice with high symmetry might have been interfered with by the local stress and distortion provided by solute Mo atoms.

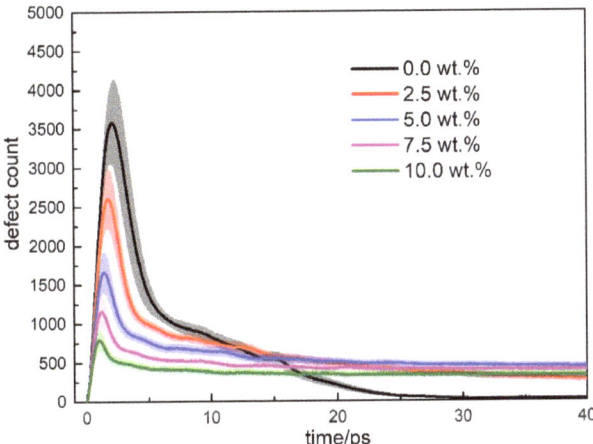

Figure 4. Time evolutions of the defect count in U-Mo systems with different Mo contents.

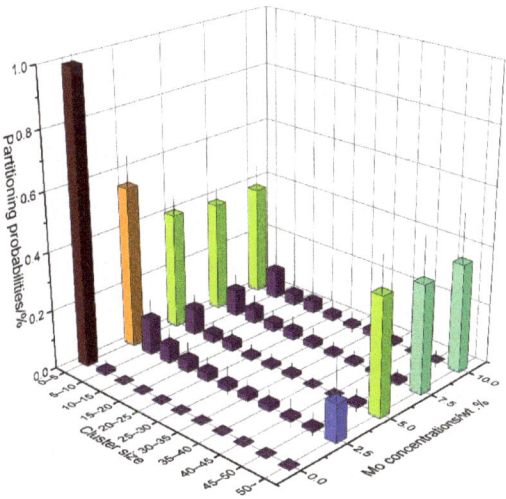

Figure 5. Size distributions of residual defect clusters versus Mo contents.

However, as shown in Figure 6a, a remarkable fall in residual defect populations was found when Mo contents exceeded 5 wt.%, suggesting that the decrease of Mo content from that in fuel alloys (>7 wt.%) brings about more irradiation damage with a potential acceleration of defect evolution. Meanwhile, the onset of IIR was suggested to be highly correlated with the accumulation of defect structures near GBs [31]. It could be further inferred that the local depletion of Mo in U-Mo potentially has an important influence on IIR and even accelerates swelling behavior from the primary stage. Actually, in U-Mo alloys with a relatively low Mo content (U-7 wt.% Mo), an increased swelling has already been observed [32].

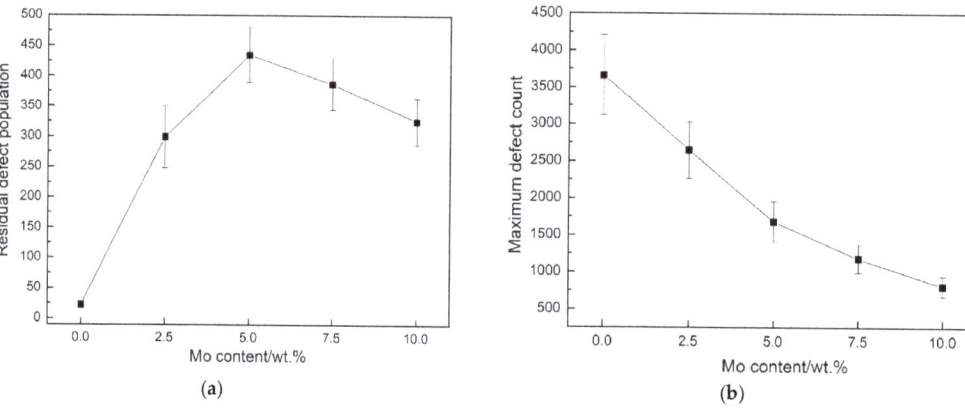

Figure 6. Statistical results of (**a**) residual defect population and (**b**) peak values of defect population during cascade simulations with different Mo contents.

4. Discussion

To investigate the underlying mechanism of the reduction of residual defect population, further attention was first paid to the spatial distribution of residual defects. Figure 7a shows the average distances between residual defects and the initial position of PKA as a function of Mo content and it could clearly be seen that the increase of Mo content restricts the spatial distribution of residual defects. Meanwhile, it could also be noted based on Figure 6b that the peak values of defect population decrease as Mo contents increase, suggesting that cascade processes are considerably limited in smaller volumes with a higher mass fraction of Mo. In general, a lattice with a lower local distortion favors a long-range transfer of kinetic energy by atomic interactions, in which case the cascade process produces more chains of interstitials squeezed in special crystal orientations, that is, crowdions. Through snapshots of defect spatial evolutions during the cascade processes, the present study also revealed that the majority of crowdions in the form of long chains vanished at the end of the simulation, which might have resulted from the reverse of displacement of those interstitial atoms. The transient formation of the crowdions could partially account the remarkable spikes in defect counts in Figure 4, especially with a lower mass fraction of Mo.

Moreover, the present study also examined the average atomic fractions of Mo among atoms with a displacement of more than 1 Å under different Mo contents, as shown in Figure 7b. Ignoring the case of pure U, Mo fractions in displaced atoms have always been observed to be lower than that in the total simulation bulk, which could also serve as a corroboration of the negative role of Mo in kinetic energy transfer. It could then be inferred that the introduction of Mo hinders the swelling of the displacement spike and thus restricts the spatial distribution of defects, which results in a higher probability of defect annihilation and a partial decrease of the residual defect population.

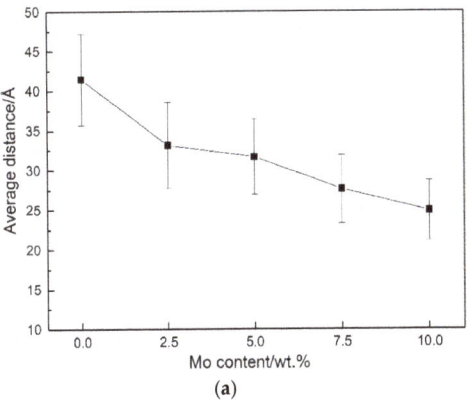 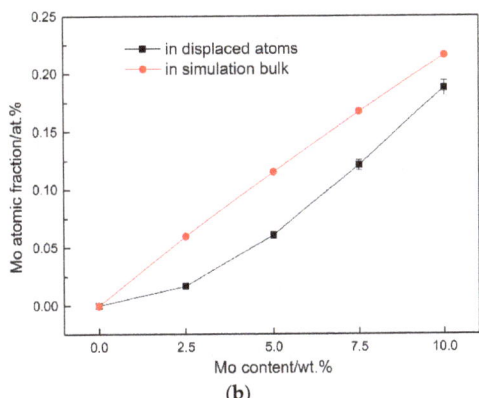

Figure 7. (a) Average distances between residual defects and the initial position of PKA as a function of Mo content and (b) Average atomic fractions of Mo among atoms with displacement more than 1 Å as a function of Mo content. The corresponding Mo atomic fractions in the total simulation bulk are also shown for comparison.

5. Conclusions

In the present research, an ADP was developed to capture atomic behaviors in U-Mo systems. Macroscopic properties including lattice constants, cohesive energies, and elastic constants were reproduced moderately well. Meanwhile, reasonable corrections were also implemented in the intermediate range of the potential, with a good agreement with DFT results. Moreover, the newly constructed potential was applied to simulations of primary radiation damage in U-Mo alloys and an uplift of residual defect population was observed when Mo content decreases to 5 wt.%. This indicates an increase of defect evolution and serves as a corroboration of the critical role of Mo depletion near GBs in the onset of IIR, and of accelerated swelling behavior in nuclear fuels.

Author Contributions: Conceptualization, W.L., J.L. (Jiahao Li), J.L. (Jianbo Liu), B.L. and W.O.; methodology, W.L. and J.L. (Jiahao Li); software, W.O.; validation, W.O.; formal analysis, W.O.; investigation, W.O.; resources, B.L.; writing—original draft preparation, W.O.; writing—review and editing, J.L. (Jianbo Liu). All authors have read and agreed to the published version of the manuscript.

Funding: This research was funded by the National Key Research and Development Program of China, grant number No. 2017YFB0702401, and the National Natural Science Foundation of China, grant number No. 51631005.

Acknowledgments: The authors acknowledge Yi Wang and Jingcheng Chen for providing inspiration for dealing with technical problems.

Conflicts of Interest: The authors declare no conflict of interest. The funders had no role in the design of the study; in the collection, analyses, or interpretation of data; in the writing of the manuscript, or in the decision to publish the results.

References

1. Kalashnikov, V.V.; Titova, V.V.; Sergeev, G.I.; Samoilov, A.G. Uranium-molybdenum alloys in reactor construction. *Sov. J. At. Energy* **1959**, *5*, 1315–1325. [CrossRef]
2. Kulcinski, G.L.; Leggett, R.D.; Hann, C.R.; Mastel, B. Fission gas induced swelling in uranium at high temperatures and pressures. *J. Nucl. Mater.* **1969**, *30*, 303–313. [CrossRef]
3. Landa, A.; Söderlind, P.; Turchi, P. Density-functional study of U–Mo and U–Zr alloys. *J. Nucl. Mater.* **2011**, *414*, 132–137. [CrossRef]
4. Kim, Y.S.; Hofman, G.L. Fission product induced swelling of U–Mo alloy fuel. *J. Nucl. Mater.* **2011**, *419*, 291–301. [CrossRef]
5. Tangri, K.; Williams, G.I. Metastable phases in the uranium molybdenum system and their origin. *J. Nucl. Mater.* **1961**, *4*, 226–233. [CrossRef]
6. Yakel, H.L. Crystal structures of transition phases formed in U/16.60 at% Nb/5.64 at% Zr alloys. *J. Nucl. Mater.* **1969**, *33*, 286–295. [CrossRef]

7. Williams, W.; Rice, F.; Robinson, A.; Meyer, M.; Rabin, B. Afip-6 mkii post-irradiation examination summary report. In *Technical Report INL/LTD-15-34142*; Idaho National Laboratory: Idaho Falls, ID, USA, 2015.
8. Kim, Y.S.; Hofman, G.L.; Cheon, J.S. Recrystallization and fission-gas-bubble swelling of U–Mo fuel. *J. Nucl. Mater.* **2013**, *436*, 14–22. [CrossRef]
9. Van den Berghe, S.; Van Renterghem, W.; Leenaers, A. Transmission electron microscopy investigation of irradiated U–7wt%Mo dispersion fuel. *J. Nucl. Mater.* **2008**, *375*, 340–346. [CrossRef]
10. Smirnova, D.E.; Kuksin, A.Y.; Starikov, S.V.; Stegailov, V.V.; Insepov, Z.; Rest, J.; Yacout, A.M. A ternary EAM interatomic potential for U–Mo alloys with xenon. *Model. Simul. Mater. Sci. Eng.* **2013**, *21*, 035011. [CrossRef]
11. Tian, X.; Xiao, H.; Tang, R.; Lu, C. Molecular dynamics simulation of displacement cascades in U–Mo alloys. *Nucl. Instrum. Methods Phys. Res. Sect. B Beam Interact. Mater. At.* **2014**, *321*, 24–29. [CrossRef]
12. Hu, S.; Setyawan, W.; Joshi, V.V.; Lavender, C.A. Atomistic simulations of thermodynamic properties of Xe gas bubbles in U10Mo fuels. *J. Nucl. Mater.* **2017**, *490*, 49–58. [CrossRef]
13. Smirnova, D.; Kuksin, A.Y.; Starikov, S. Investigation of point defects diffusion in bcc uranium and U–Mo alloys. *J. Nucl. Mater.* **2015**, *458*, 304–311. [CrossRef]
14. Smirnova, D.E.; Kuksin, A.Y.; Starikov, S.V.; Stegailov, V.V. Atomistic modeling of the self-diffusion in γ-U and γ-U-Mo. *Phys. Met. Metallogr.* **2015**, *116*, 445–455. [CrossRef]
15. Starikov, S.; Kolotova, L.; Kuksin, A.Y.; Smirnova, D.; Tseplyaev, V. Atomistic simulation of cubic and tetragonal phases of U-Mo alloy: Structure and thermodynamic properties. *J. Nucl. Mater.* **2018**, *499*, 451–463. [CrossRef]
16. Chen, J.; Ouyang, W.; Lai, W.; Li, J.; Zhang, Z. A new type angular-dependent interatomic potential and its application to model displacement cascades in uranium. *Nucl. Instrum. Methods Phys. Res. Sect. B Beam Interact. Mater. At.* **2019**, *451*, 32–37. [CrossRef]
17. Ziegler, J.F.; Biersack, J.P. *The Stopping and Range of Ions in Matter*; Springer: Boston, MA, USA, 1985; pp. 93–129.
18. Stoller, R.E.; Tamm, A.; Beland, L.K.; Samolyuk, G.D.; Stocks, G.M.; Caro, A.; Slipchenko, L.V.; Osetsky, Y.N.; Aabloo, A.; Klintenberg, M.; et al. Impact of Short-Range Forces on Defect Production from High-Energy Collisions. *J. Chem. Theory Comput.* **2016**, *12*, 2871–2879. [CrossRef] [PubMed]
19. Kresse, G.; Furthmüller, J. Efficiency of ab-initio total energy calculations for metals and semiconductors using a plane-wave basis set. *Comput. Mater. Sci.* **1996**, *6*, 15–50. [CrossRef]
20. Plimpton, S. Fast Parallel Algorithms for Short-Range Molecular Dynamics. *J. Comput. Phys.* **1995**, *117*, 1–19. [CrossRef]
21. Stukowski, A. Visualization and analysis of atomistic simulation data with OVITO–the Open Visualization Tool. *Model. Simul. Mater. Sci. Eng.* **2009**, *18*, 015012. [CrossRef]
22. Fernández, J.R.; Pascuet, M.I. On the accurate description of uranium metallic phases: A MEAM interatomic potential approach. *Model. Simul. Mater. Sci. Eng.* **2014**, *22*, 055019. [CrossRef]
23. Kittel, C.; McEuen, P.; McEuen, P. *Introduction to Solid State Physics*; Wiley: Hoboken, NJ, USA, 1996; Volume 8.
24. Barrett, C.S.; Mueller, M.H.; Hitterman, R.L. Crystal Structure Variations in Alpha Uranium at Low Temperatures. *Phys. Rev.* **1963**, *129*, 625–629. [CrossRef]
25. Yoo, C.-S.; Cynn, H.; Söderlind, P. Phase diagram of uranium at high pressures and temperatures. *Phys. Rev. B* **1998**, *57*, 10359–10362. [CrossRef]
26. Taylor, C.D. Erratum: Evaluation of first-principles techniques for obtaining materials parameters of alpha-uranium and the (001) alpha-uranium surface. *Phys. Rev. B* **2009**, *80*, 149906. [CrossRef]
27. Wilson, A.S.; Rundle, R.E. The structures of uranium metal. *Acta Crystallogr.* **1949**, *2*, 126–127. [CrossRef]
28. Edwards, J.W.; Speiser, R.; Johnston, H.L. High Temperature Structure and Thermal Expansion of Some Metals as Determined by X-Ray Diffraction Data. I. Platinum, Tantalum, Niobium, and Molybdenum. *J. Appl. Phys.* **1951**, *22*, 424–428. [CrossRef]
29. Errandonea, D.; Schwager, B.; Ditz, R.; Gessmann, C.; Boehler, R.; Ross, M. Systematics of transition-metal melting. *Phys. Rev. B* **2001**, *63*, 132104. [CrossRef]
30. Stearns, M.B. Landolt-Bornstein: Numerical Data and Functional Relationships in Science and Technology, Group III Condensed Matter. *At. Defects Metals.* **1991**, *25*. [CrossRef]
31. Frazier, W.E.; Hu, S.; Burkes, D.E.; Beeler, B.W. A Monte Carlo model of irradiation-induced recrystallization in polycrystalline UMo fuels. *J. Nucl. Mater.* **2019**, *524*, 164–176. [CrossRef]
32. Van Den Berghe, S.; Lemoine, P. Review of 15 years of high-density low-enriched umo dispersion fuel development for research reactors in europe. *Nucl. Eng. Technol.* **2014**, *46*, 125–146. [CrossRef]

Review

Strain Rate Effect on the Thermomechanical Behavior of NiTi Shape Memory Alloys: A Literature Review

Zhengxiong Wang, Jiangyi Luo, Wangwang Kuang, Mingjiang Jin, Guisen Liu, Xuejun Jin * and Yao Shen *

The State Key Laboratory of Metal Matrix Composites, School of Materials Science and Engineering, Shanghai Jiao Tong University, 800 Dongchuan Road, Shanghai 200240, China
* Correspondence: jin@sjtu.edu.cn (X.J.); yaoshen@sjtu.edu.cn (Y.S.)

Abstract: A review of experiments and models for the strain rate effect of NiTi Shape Memory Alloys (SMAs) is presented in this paper. Experimental observations on the rate-dependent properties, such as stress responses, temperature evolutions, and phase nucleation and propagation, under uniaxial loads are classified and summarized based on the strain rate values. The strain rates are divided into five ranges and in each range the deformation mechanism is unique. For comparison, results under other loading modes are also reviewed; however, these are shorter in length due to a limited number of experiments. A brief discussion on the influences of the microstructure on the strain-rate responses is followed. Modeling the rate-dependent behaviors of NiTi SMAs focuses on incorporating the physical origins in the constitutive relationship. Thermal source models are the key rate-dependent constitutive models under quasi-static loading to account for the self-heating mechanism. Thermal kinetic models, evolving from thermal source models, address the kinetic relationship in dynamic deformation.

Keywords: NiTi; shape memory alloy; strain rate effect; thermomechanical

1. Introduction

NiTi Shape Memory Alloys (SMAs) are the most widely used in the shape memory material family, and have important applications in electronic components, medical devices, shock absorption devices, etc., thanks to their unique thermomechanical behaviors [1–4]. Reliable and predictive simulations of NiTi SMA wires under high strain rates are always required for actuation and auxiliary control. An example of a hybrid device with SMA wires is shown in Figure 1, which is designed for anti-seismic structure to restrict the strain within 6% [5]. These growing demands push investigations on both quasi-static and dynamic responses of NiTi SMAs [6–10].

Researchers have conducted experiments on NiTi SMAs at a wide range of strain rates and most experiments have been under uniaxial loads [11]. While the SMA specimens could be bars, wires, or plates depending on the specific experiments, general observations and mechanisms could be extracted and summarized regardless of the shape and size of specimens. The stress responses and temperature evolutions during transformation have varied depending on the strain rate. At a very low strain rate $\dot{\varepsilon} < 10^{-4}\text{s}^{-1}$, the stress and temperature scarcely changed since the latent heat dissipated to surroundings sufficiently [11–13]. As the strain rate rises to $10^{-4}\text{s}^{-1} < \dot{\varepsilon} < 10^{-1}\text{s}^{-1}$, the stress and temperature increases with increasing strain rate because part of the heat by transformation is left in the material [11,14–16]. The heat exchange rate between the specimen and the environment is found to be greatly influenced by the temperature rise in this strain rate range, as demonstrated in a comparison test by Shaw and Kyriakides [11] between water-enclosed and air-enclosed NiTi SMA wires. In the medium range of strain rate $10^{-1}\text{s}^{-1} < \dot{\varepsilon} < 10^{3}\text{s}^{-1}$, the stress stopped increasing but the temperature increased quickly since an adiabatic process was reached [17–21]. A quickly-recovered residual strain was reported by Chen et al. [17] around 81~750 s^{-1}, which was ascribed to the significant thermal

hysteresis as a result of an adiabatic process. In strain rates between $10^3 s^{-1} < \dot{\epsilon} < 10^4 s^{-1}$, a sudden rise of the critical transformation stress takes place, and this can be attributed to the drag effect of dislocations surrounding the phase interface [22]. Nemat-Nasser et al. [23–25] observed considerably increased rate sensitivity and curved interfaces of martensite in TEM micrographs. The final strain rate range was $\dot{\epsilon} > 10^4 s^{-1}$, in which plastic deformation occurs in austenite without martensitic transformation because martensitic transformation no longer satisfies the demand of quick deformation [24,26–28].

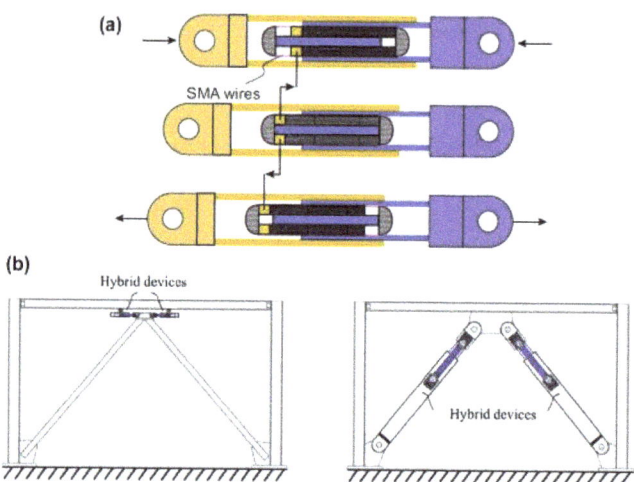

Figure 1. (a) Hybrid device with SMA wires and (b) different configurations of the proposed device [5]. (Reprinted from Ref. [5], Figure 21.1, 2021, with permission from Elsevier.)

Furthermore, the nucleation and propagation of phase transformation under various strain rates have also been investigated by many researchers [11,29,30]. When the strain rate is very low, the martensite nucleates at a few locations and then extends to all fields [11,29,31]. As the strain rate increases, the number of nucleation locations grows, and martensitic transformation domains propagate in a parallel mode rather than entirely [30,31].

Compared to uniaxial tests, experiments under shear and indentation have been conducted less frequently [32]. Simple shear tests on NiTi SMAs were performed early by Manach and Favier [33], while a more comprehensive study within a wider range of strain rates from 10^{-4} s^{-1} to 10^3 s^{-1} was accomplished by Huang et al. [34]. As for indentation tests, Amin et al. [35] and Shahirnia et al. [36] both pointed out that the maximum indentation was influenced by the loading rate. In cases of cyclic loading, superelasticity degeneration and temperature variations have been found to be strongly dependent on the strain rate [37,38].

Besides the loading mode, microstructure also has an important impact on the rate-dependent behaviors of NiTi SMAs [25,39–41]. The influences of R-phase, precipitated phase, and grain size on the strain rate responses of general SMAs have been explored in many studies [39,40,42,43]. The influence of R-phase transformation on the strain effect sensitivity has been investigated by Helbert et al. [39] on NiTi SMA wires. The precipitation evolution of Ni_4Ti_3 and transformation behaviors have been studied and characterized by Fan et al. [42] in quasi-static loading and Yu et al. [43] in impact loading. The grain size influenced the amount of transformable martensite and heat generation, and therefore influenced the temperature field and transformation stress [40]. For porous and composite SMAs, higher strain rates brought in a greater transformation stress similar to general NiTi SMAs [25,41].

Modeling the strain rate effect of NiTi SMAs at low and medium strain rates is focused on characterizing and qualifying the thermal effect during transformation, where the latent heat plays an important role. Thermal source models were proposed to represent the strain rate effect as extra heat sources added to the energy equation [44–47]. The thermodynamic potential is usually adopted in thermal source models to derive the temperature evolution equation, and can be obtained either empirically or theoretically [48,49]. Many simulations have been performed at strain rates ranging from very low to medium on the basis of the thermal source models, and their predictions have matched well with experiments at corresponding strain rates [50–55]. However, the drag effect surpasses the thermal effect at high strain rates. By considering the kinetic effect and adding a resistance force into the constitutive relationship, thermal kinetic models have been extended from thermal source models to describe the thermomechanical behaviors of NiTi SMAs at high strain rates [56–58].

Multiple studies have conducted experiments and developed models for the deformation behaviors of NiTi SMAs under various strain rates. However, there is hardly any systematic analysis of the strain rate effect, except for some subsections touching on the strain rate effect in general reviews for NiTi SMA microstructure evolution and thermodynamic behaviors [59–61]. Therefore, this paper aims to present a comprehensive review and summarize the common physical mechanisms of the strain rate effects in a wide strain-rate range from quasi-static to dynamic loading. The strain rate effect under uniaxial loads will be elaborated in the first place since the number of uniaxial tests is the largest. The rate effect under other loading modes, such as shear, indentation, and cyclic loading, will be discussed later. A brief summary of the dependence of the strain rate effect on microstructure is presented at the end of Section 2. Section 3 recapitulates the main constitutive models capturing the strain rate effect. The approach is to construct the thermal source model which will first be introduced and then followed by a discussion of thermal-source components. Thermal kinetic models will be discussed next. Simulation examples will be given in both models. Final remarks with future research directions are given in Section 4.

2. Experimental Observations

Great progress has been made in uniaxial experiments to investigate the pseudo-elastic and yielding behaviors of NiTi SMAs under different strain rates. Shear and indentation tests have also been conducted, though less frequently compared to uniaxial. Experimental results under different loading rates have shown that the critical transformation stress increases as the strain rate grows. In addition to the loading mode, microstructure is another factor to influence the strain rate effect, which will be discussed at the end of this section.

2.1. Strain Rate Effect under Uniaxial Loads

The uniaxial experimental tests have been conducted by either tension or compression. Tensile test experiments are more common and frequent at low and medium loading rates, while the quantity of compression tests under shock conditions with the split Hopkinson bar are more numerous [18,26,32,62]. Compared to those in tension, stress-strain curves in compression tests have less recoverable strain, a steeper transformation slope, and higher critical stress. Tension-compression asymmetry is strongly dependent on the level of strain rate and temperature [63,64].

Under the uniaxial loading mode, the critical transformation stress is an important indicator of the strain rate effect on NiTi SMAs. To better explain the rate effects on the transformation stress over a wide range of strain rates, the transformation stress is constructed as a function of the strain rate by this review in Figure 2, based on the previous work by Nemat-Nasser et al. [23–25,65] and Zurbitu et al. [20,21]. On the whole, at a temperature below the maximum temperature for stress-induced martensite formation, the martensitic transformation stress (σ_t) and the austenite yield stress (σ_y) increase with strain rate. Specifically, at $\dot{\epsilon} < 10^4 \, s^{-1}$, stress-induced martensitic transformation always happens devoid of austenite yielding as $\sigma_y > \sigma_t$, where σ_y can be obtained using tensile experiments under high temperature. In shock conditions where $\dot{\epsilon} > 10^4$ s^{-1}, austenite yields without any martensitic transformation as now $\sigma_y < \sigma_t$.

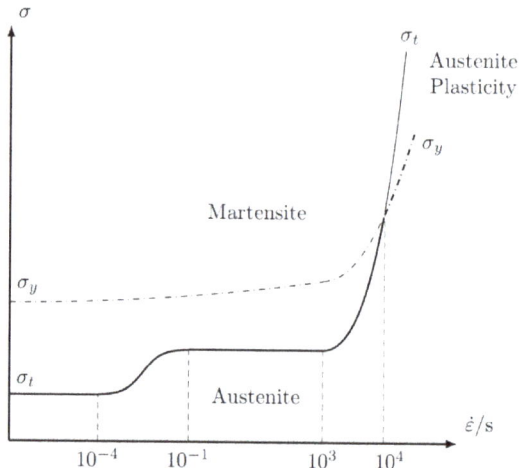

Figure 2. A general plot of the martensitic transformation stress (σ_t) and the austenite yield stress (σ_y) as functions of the strain rate.

The features for the deformation mechanisms of NiTi SMAs at various strain rates can be classified into categories in five strain ranges separated by four critical strain rate values, which are 10^{-4} s^{-1}, 10^1 s^{-1}, 10^3 s^{-1}, and 10^4 s^{-1}.

- Basically, at a very low strain rate $\dot{\epsilon} < 10^{-4}$ s^{-1}, phase transformation heat dissipates sufficiently to the environment, thus the temperature change of the material can be ignored [11–13]. The martensitic transformation stress is therefore not sensitive to the strain rate in this isothermal process, which is shown as a stress platform in Figure 2.
- In a relatively low strain range, 10^{-4} s$^{-1} < \dot{\epsilon} < 10^{-1}$ s^{-1}, the influence of the latent heat, dissipation energy, and elastic heat gradually grows, leading to small temperature variations of the material [11,14–16]. The strain rate influences the temperature field and brings an increase in the martensitic transformation stress.
- As the strain rate climbs to a medium range of 10^{-1} s$^{-1} < \dot{\epsilon} < 10^3$ s^{-1}, the transformation stress is nearly unchanged [17–21]. This is because the temperature evolution is insensitive to the strain rate as the deformation process is adiabatic. The heat produced, with amount proportional to the volume of transformed martensite, is totally used to warm up the sample, while the rate-sensitive heat produced by the transformation-induced plasticity strain is negligible.
- A sudden rise of the transformation stress appears at strain rates between 10^3 s$^{-1} < \dot{\epsilon} < 10^4$ s^{-1}, where the dislocation drag effect becomes more significant and the flow stress is more sensitive to the strain rate [22–25,65–69]. The overall stress level increases remarkably as well as the transformation stress.
- The final strain rate region is $\dot{\epsilon} > 10^4$ s^{-1}, in which martensitic transformation cannot provide enough deformation, thus parent austenitic phase plastic deformation occurs due to the extremely high strain rate [24,26–28].

Experimental results and deformation mechanisms in all these five strain-rate ranges will be elaborated individually in the next section.

2.1.1. Less Than 10^{-4} s^{-1}

Isothermal processes are always observed in NiTi SMA experiments under very low loading rates. This is because the heat energy generated by transformation dissipated sufficiently into the environment [11–13]. In the last century, researchers measured the strain rate threshold, below which the sample temperature and transformation stress only changed in a negligible small range [11–13]. The threshold value is finally given around 10^{-4} s^{-1}.

According to the experimental results obtained by Shaw and Kyriakides [11], martensite nucleates at few locations and subsequently extends to all NiTi crystallites when the strain rate is below 10^{-4} s^{-1}. They installed four strain sensors on the wire to monitor the nuclear situation of the martensite during transformation in 70 °C water. The locations of four sensors are shown in Figure 3.

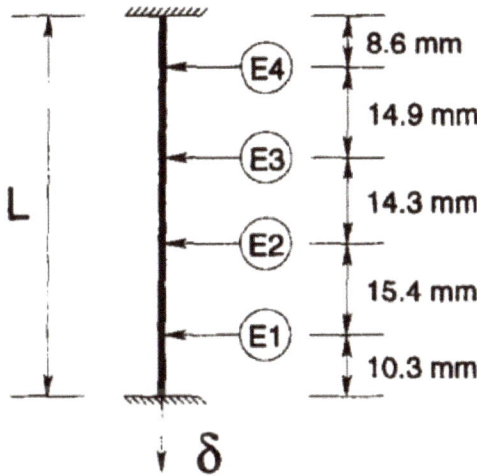

Figure 3. Photograph of a NiTi wire specimen with four extensometers [11]. (Reprinted from Ref. [11], Figure 8, 1995, with permission from Elsevier.)

Based on the temperature, stress, and strain histories obtained with four sensors (Figure 4a), the propagation path can be captured by connecting the transformation points between austenite and martensite along the time, as shown in Figure 4b. The stress level was not sufficient to drive multiple phase fronts in this very slow loading test; therefore, only one front propagates from one side to the other during loading and unloading. The symbol A and M represent, respectively, the region of austenite and martensite. Note that small temperature jumps were recorded (depicted with T1 squares in Figure 4a) when the A/M interface crossed the thermocouple at the middle point of the wire. The two temperature jumps peak during loading and valley during unloading, indicating rapid local loss and gain of heat, respectively, which confirms the isothermal hypothesis. It can be seen from Figure 4a that the stress roughly changes with the strain in both forward and reverse transformation under this very low strain rate.

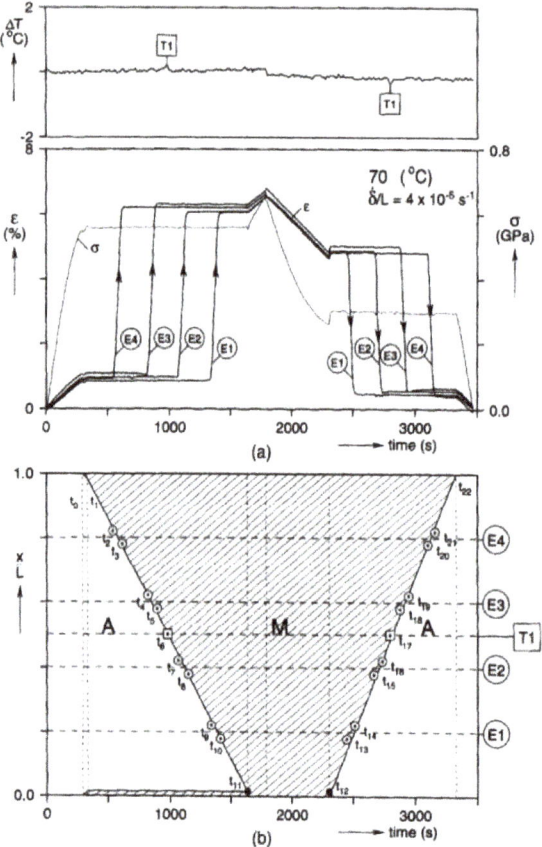

Figure 4. (a) Temperature, stress, and strain histories at four sensor positions under a strain rate of $4 \times 10^{-5} \mathrm{s}^{-1}$; (b) Location-time diagram of significant events [11]. (Reprinted from Ref. [11], Figure 11, 1995, with permission from Elsevier.)

2.1.2. From 10^{-4} s^{-1} to 10^{-1} s^{-1}

As the loading rate grows, the heat energy generated during transformation, such as the latent heat, dissipation heat, and thermal expansion heat, cannot be released into the environment sufficiently. Excessive heat will lead to a temperature rise in the material, which in turn influences the transformation rate. The critical transformation stress typically increases with the strain rate; however, the environment temperature and heat exchange rate also play crucial roles in this thermal-mechanical coupling mechanism. Among the studies, Shaw and Kyriakides [11] compared the tensile properties between water-enclosed and air-enclosed NiTi SMA wires. Their experiments showed that the wire temperature varied remarkably between two surroundings and they found that the transformation stress in the water-enclosed wire was much lower than that of the air-enclosed one. The strain rate effect is weaker in the water-enclosed wire considering a higher efficiency of heat exchange. Bruhns [15] and Grabe's group [16] spent some effort on decoupling the thermal effect from strain-rate viscous effects at high temperatures and came to the conclusion that for a specified temperature range NiTi SMA wires can be seen as deformation-rate independent materials.

The number of martensite nucleation sites increases with increasing strain rate, and these martensite domains propagate in a parallel mode. Gadaj et al. [29] applied a thermovision camera to record the infrared radiation emitted by the specimen surface. They observed the number of martensite domains became greater as the strain rate increased, and these domains propagated parallel to the austenite strip. Similar nucleation and propagation modes were observed by Zhang et al. [31].

Tobushi et al. [14] investigated the deformation behaviors of NiTi SMA under strain-controlled and stress-controlled conditions. At a low strain rate in strain-controlled situations, they pointed out a stress overshoot at the temperature Ms and a stress undershoot at the temperature As, respectively. In contrast, the overall stress-strain curves subjected to the stress control are similar to those in strain-controlled cases with high strain rate. While at a high strain rate, stress-controlled experimental results have exhibited a smooth transition in stress around the transformation temperature M_s and A_s. The difference can be explained by the excessive energy needed for nucleation when the phase interface starts moving, since greater stress is needed to transform during strain-controlled situations.

In the case of compression tests, the transformation stress also increases with increasing strain rate; however, the transformation mode does not change with the strain rate from 10^{-4} s^{-1} to 10^3 s^{-1} and the strain field is generally more uniform compared with tension tests. In contrast to the localized nucleation and propagation of martensite bands under tensile loading, NiTi SMA subjected to compressive loading always exhibits a more homogeneous transformation. Elibol and Wagner [64] employed a digital image correlation in situ technique to show that the surface strain fields in compression were always uniform during transformation without any strain localization. Meanwhile, the transformation mode was barely influenced by an increase of the strain rate in both quasi-static and dynamic conditions.

The finishing point of the martensitic transformation becomes harder to distinguish when the strain rate is higher. Dayananda and Rao [70] have tested NiTi SMA wires at strain rates from 3.3×10^{-5} s^{-1} to 3.0×10^{-2} s^{-1}. Three "elastic" segments were identified in the stress-strain curves when the strain rate was low, which were the elastic austenite segment, superelasticity segment, and elastic Stress-Induced Martensite (SIM) segment. As the strain rate increased above 5.0×10^{-3} s^{-1}, a fourth segment emerged between the superelasticity and elastic stress-induced martensite segments. This intermediate segment resulted from the overlapping of the SIM formation and elastic SIM deformation, and its length increased with increasing strain rate.

2.1.3. From 10^{-1} s^{-1} to 10^3 s^{-1}

An adiabatic process happens under this medium strain rate. The energy generated by phase transformation accumulates in the transformed region and cannot be released in such a short time. The temperature of the transformed region increases dramatically; however, the stress level remains the same. This is because the total transformed heat production is finite.

The range of strain rate from 10^2 s^{-1} to 10^3 s^{-1} can be reached by the split Hopkinson bar. Experiments conducted by Chen et al. [17] showed that when the strain rate came to 81~750 s^{-1}, there would be a residual deformation in the specimen, which recovered in a few seconds to several hours. This residual deformation can be attributed to the thermal hysteresis as a result of the adiabatic process. Another paper published by Chen and Bo [62] in 2006 paid close attention to the unload progress of split Hopkinson bar systems. They reached a stable strain rate around 430 s^{-1} under both loading and unloading conditions, which used to have difficulties in dynamic experiment design. Nemat-Nasser and Choi [18] also found that the initial temperature affected the deformation mechanism when strain rates came to 500~700 s^{-1}. The transformation stress increased with the initial temperature, and austenite plastic deformation happened eventually. Regarding to the compression test, Adharapurapu et al. [63] discovered that the asymmetry between compression and tension became weaker with higher temperature, and this phenomenon was more conspicuous under high strain rates, as shown in Figure 5. Recently, Shen et al. [19] focused on the thermal evolution in this medium strain rate range by the split Hopkinson bar system and an infrared detection system. They compared martensite NiTi SMA wires with superelastic wires between the transformation temperature A_s and A_f, and pointed out that martensite wires had higher dissipated energy than superelastic ones. This could be explained by larger plastic deformation observed in martensite wires.

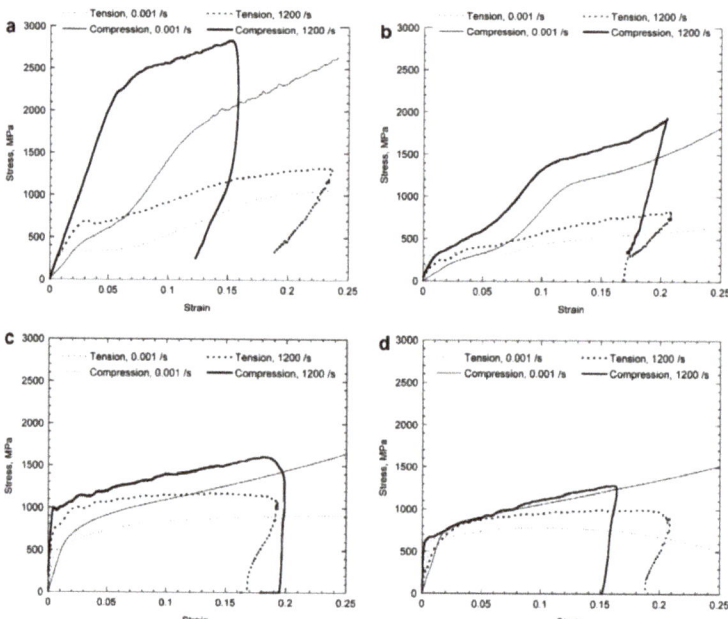

Figure 5. Compression and tension stress-strain curves of NiTi SMA. The effect of temperature on compression-tension asymmetry at dynamic (10^3 s^{-1}) and quasi-static (10^{-3} s^{-1}) strain rates. (a–d) are at −196 °C, 0 °C, 200 °C, and 400 °C, respectively [63]. (Reprinted from Ref. [63], Figure 5, 2006, with permission from Elsevier.)

At medium strain rates, nucleation sites are too numerous to distinguish but merge into large martensite zones. As observed by Saletti et al. [30] at a strain rate round 20 s^{-1}, two large martensite zones emerge on the ends of the bar specimen (one at each end) where the loading is applied, and the length of transformed martensite enlarged with two fronts move in opposite direction towards the specimen center. However, the details of the localized phase nucleation and propagation are still obscure due to the experimental difficulty [30].

However, methods are lacking to test material properties around 10^{-1}~10^2 s^{-1}, where car crashes and gravity-dropped bombs are the typical examples. Conventional mechanical test methods such as tension, compression, and bending test performed with screw-driven or servohydraulic load-frames are only applicable below this strain rate range. The split-Hopkinson bar technique has permitted the evaluation of mechanical properties over durations shorter than a millisecond, which is higher than this rate range. In another word, testing at intermediate rates is inherently challenging due to the possibility of elastic wave reflections and the difficulty in establishing dynamic equilibrium in the sample and the load sensors [71].

This problem can be partly solved by improving experimental devices. Xu et al. [72] used an impact testing system for the first time to show the influence of temperature and impact velocity. However, the device cannot control the strain rate precisely. Zurbitu et al. [20,21] explored NiTi SMA wire properties on the order of 1~10^2 s^{-1} using an instrumented tensile-impact technique. Different from low strain rates, medium strain rates led to more stable critical transformation stresses for both austinite and martensite phases. Zurbitu's paper compared the critical transformation stress results with those at low strain rates and demonstrated the adiabatic process during stretching in SMA wires.

2.1.4. From 10^3 s^{-1} to 10^4 s^{-1}

The critical transformation stress increases remarkably with the strain rate in this range. The rise of transformation stress can be explained by the dislocation drag mechanism in the plastic deformation around the martensite interface [73,74]. Dislocation slips at very high strain rates need a much higher driving stress compared to those at the low strain rate due to the phono drag effect [60]. The phase interface moves with the high-speed dislocations on it and is thus subjected to the dislocation drag effect as well. Since the velocity of dislocations is a key factor in shock dynamics, the velocity of martensite interfaces should be considered in the transformation mechanism at high strain rates [23,65,74].

Split-Hopkinson bar systems are commonly used to conduct experimental tests in this strain rate range. Hudspeth et al. [75] developed a new technique to investigate the dynamic behaviors of SMA materials through simultaneous X-ray imaging and diffraction and gained a strain rate of 5000 s^{-1}. Nemat-Nasser et al. [23–25] improved the Split-Hopkinson bar systems at a high strain rate in 2005. They showed that strain rate sensitivity increased sharply and observed a curved interface of martensite in TEM micrographs. Therefore, the velocity of the martensite interface migration plays an important role at this stage. Yang et al. [28] used a dynamic ex-situ neutron diffraction technique to characterize the rate effect of NiTi SMAs. Based on Yang's results, as the strain rate increases, plastic deformation replaces the martensite reorientation and the volume fraction of detwinning martensite decreases.

Recently, several Molecular-Dynamics (MD) models have been built to analyze the deformation mechanism of NiTi SMAs at a high strain rate level of around 10^3 s^{-1}. For instance, Wang et al. [76] and Yazdandoost et al. [77] studied SMA crystallographic structure change in the transformation progress with MD in the shock condition. Yin et al. [78] and Ko et al. [79] simulated a NiTi nanopillar with MD under various strain rates and temperatures. They showed that the phase stress of B19 $\rightarrow$ B19$'$ increased at a high strain rate because there was not enough time for atoms to reach new positions. Yazdandoost et al. [80] focused on the dissipation energy in the shock condition and indicated that transformation provided the main dissipative function.

2.1.5. Greater Than 10^4 s^{-1}

When the strain rate reaches 7000 s^{-1}, plastic deformation of austenite occurs without any martensitic transformation [27,28], which implies the yielding stress for austenite rises with the strain rate more slowly than that of the transformation stress. In the meantime, Nemat-Nasser and Choi [24] found dislocation-induced plastic slips in the austenite phase with TEM, and a similar situation was observed in NiTiCr. In other words, plastic deformation takes place before the phase interface moves, thus martensite cannot emerge at this very high strain rate. Zhang et al. [26] tested NiTi alloys in the range from 10^4 s^{-1} to 10^7 s^{-1} by the technique of magnetically driven quasi-isentropic compression and shock compression. They discovered that the elastic limit increased dramatically compared to that at the low strain rate. Some other shock tests with very high strain rates [81] are beyond the scope of this paper.

2.2. Strain Rate Effect in Different Loading Modes

Besides the uniaxial loading mode, it is common for NiTi SMAs to serve in complex mechanical loading modes including shear and indentation. The number of shear and indentation tests is much smaller than that of uniaxial ones. In general, the critical transformation stress increases with increasing strain rate for all loading types. In cases of cyclic loading, superelasticity degeneration and temperature variations become more remarkable as the strain rate grows.

2.2.1. Shear

A regular transformation hardening is observed under shear loads, and the hardening is enhanced by increasing the strain rate. Simple shear experiments on NiTi SMAs were early performed by Manach and Favier [33] in quasi-static conditions. Later, a more comprehensive study on NiTi SMA behaviors under shear stress was conducted by Huang et al. [34] over a large range of strain rates from 10^{-4} s^{-1} to 10^3 s^{-1}. The low and intermediate loading rates were realized on a modified MTS machine, while the impact loading rate was achieved by Split Hopkinson bars. Heterogeneous strain fields and increased transformation stress were found in all three strain-rate conditions. A shear band was observed in the 10° direction of shear loading at low and intermediate strain rates, as shown in Figure 6a, while two separated bands emerge at the impact strain rate of 290 s^{-1}, as shown in Figure 6b. Apart from the strain localization, the thermomechanical behaviors of NiTi SMA under shear loading was similar to those under tensile loading.

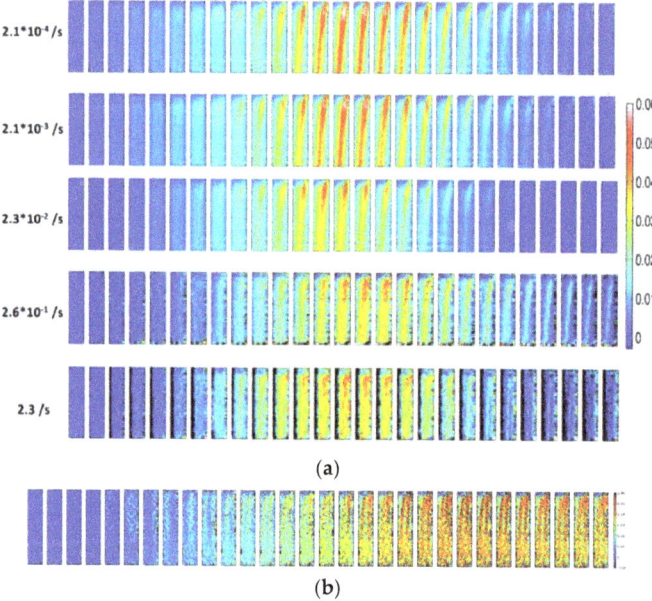

Figure 6. (a) Shear strain contours with shear bands evolutions at five different strain rates from 10^{-4} s^{-1} to 10^1 s^{-1}; (b) separated bands at a strain rate of 10^2 s^{-1} [34]. (Adapted from Ref. [34], Figures 10 and 18, 2017, with permission from Elsevier.)

2.2.2. Indentation

Similar to the strain rate effect under uniaxial loading, an increase in loading rate in indentation tests leads to a rise of transformation stress. Moreover, a higher loading rate brings a smaller indentation depth and a drop in the recoverable deformation [36]. The interpretation in most studies [35,36,82] is that the underlying mechanism can be attributed to the increased release rate of latent heat during transformation with increasing strain rate and the strong temperature dependence of NiTi SMAs.

Considering the effect of the released latent heat conduction within the material, Amini et al. [35] adopted a normalized loading rate parameter to represent the rate effect in the experiment. The normalized loading rate was defined as a ratio of the loading rate $\dot{F}$ to the transformation stress σ_{y0} times the conduction coefficient k, and the normalized indentation depth was defined as a ratio of the indentation depth h to the radius of the tip R, as shown in Figure 7. Based on the level of the normalized loading rate, the indentation process could be roughly classified into isothermal, adiabatic, and a transition between them, which resembles the strain-rate-dependent ranges under uniaxial loads. Amini et al. compared NiTi SMA with copper and quartz showing that the indentation depth of NiTi SMA decreased more rapidly with the loading rate. This can be explained by the enhanced transformation hardening caused by the latent heat accumulation around the indentation.

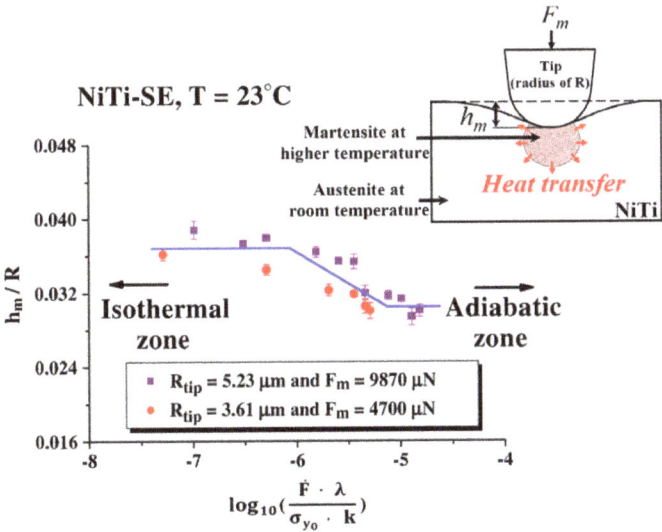

Figure 7. The rate dependence of the normalized indentation depth on the normalized loading rate parameter. A sketch of the heat transfer during indentation is drawn at the top right corner [35]. (Reprinted from Ref. [35], Figure 3, 2011, with permission from Elsevier.)

However, Farhat et al. [82] developed a simple heat model to predict the impact of temperature in indentation and suggested that the decrease of the indentation depth was not because of the temperature accumulation during transformation. Farhat et al. argued that the indentation depth decreased with the strain rate even at extremely low loading rates, where the generated heat could hardly accumulate. The loss of superelasticity might be due to the retardation of the transformation process during indentation, though more studies are needed to prove this claim.

2.2.3. Cyclic Loading

The thermomechanical cyclic behaviors of NiTi SMA have been widely studied under strain- or stress-controlled uniaxial loading mode [37,38,83–85]. The rate-dependence under quasi-static cyclic strain-controlled loadings was systematically investigated by Kan et al. [37]. Impact fatigue tests were conducted less frequently than quasi-static ones [38]. Superelasticity degeneration and temperature variations were found strongly dependent on the strain rate, while the cyclic transformation path did not changed with strain rate.

Superelasticity degeneration indicated by transformation stress decrease, residual strain accumulation, and hysteresis loss takes place at all strain rates. Generally, the start and peak of transformation stress decrease with increasing number of cycles [37,38,83]. In Kan's experiments, cyclic tests were performed at six strain rates ranging from 10^{-4} s^{-1} to 10^{-2} s^{-1} with the maximum strain fixed at 9%. The corresponding stress-strain curves in different cycles are shown in Figure 8. The drop of transformation stress at each cycle becomes more conspicuous with increasing strain rate. Residual strain accumulation during cyclic loadings increases remarkably with increasing strain rate. The dissipation energy, i.e., the area of the stress-strain hysteresis loop, decreases with cycles, but the rate-dependence of the dissipation energy loss could be more complicated.

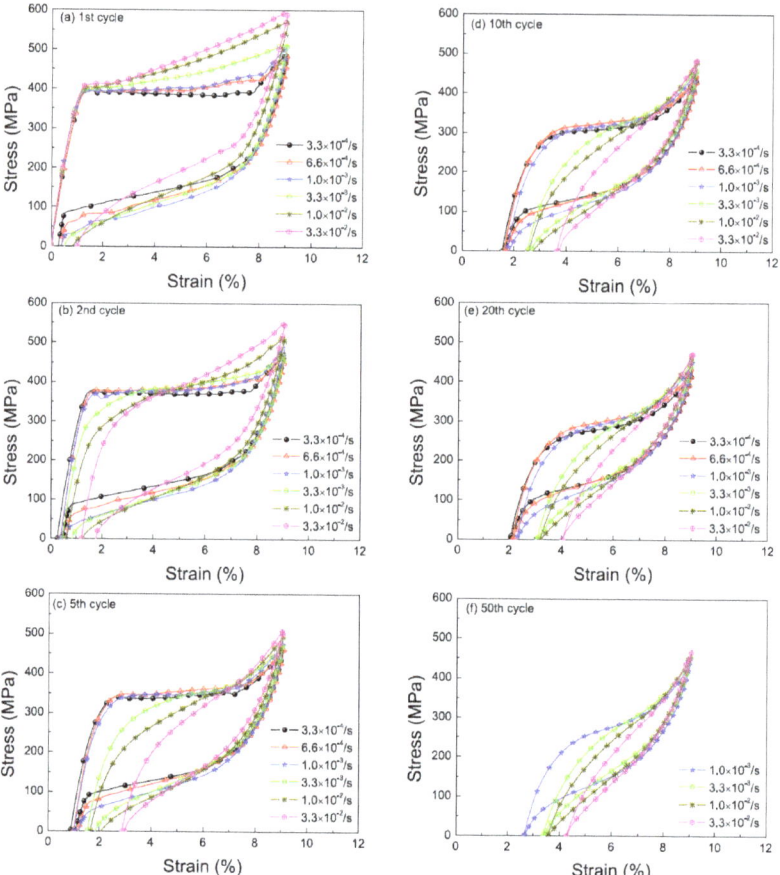

Figure 8. Rate-dependent stress–strain curves in different cycles: (**a**) 1st cycle; (**b**) 2nd cycle; (**c**) 5th cycle; (**d**) 10th cycle; (**e**) 20th cycle; (**f**) 50th cycle [37]. (Reprinted from Ref. [37], Figure 3, 2016, with permission from Elsevier.)

Superelasticity degeneration of NiTi SMA is mainly attributed to the interactions between transformation and dislocations [37,84,85]. Dislocations can be nucleated by high local stress near the phase interface and accumulate with the loading cycles. The internal stress caused by the dislocations assists stress-induced martensitic transformation, accounting for a decreasing critical transformation stress, and hindering the reverse martensitic transformation, resulting in an increasing residual strain. Therefore, superelasticity degeneration is speeded up by high strain rates.

Temperature variations during the cyclic deformation are also greatly influenced by the strain rate [37,84,85]. The evolution of temperature at five different strain rates in the 1st and 20th cycle are shown in Figure 9. In the loading part the temperature increases due to the release of transformation latent heat, while in the unloading part the temperature decreases first since the latent heat is absorbed in the reverse transformation and finally returns to the initial ambient temperature. Higher strain rates bring higher average temperatures. Generally, the amplitude of temperature oscillation decreases with increasing number of cycles; however, it increases with increasing strain rate.

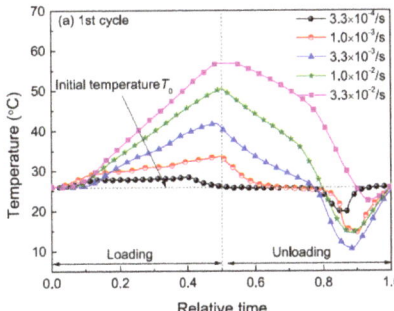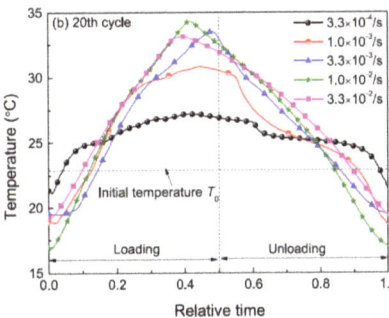

Figure 9. Temperature records at various strain rates in the 1st cycle (**a**) and 20th cycle (**b**) [37]. (Reprinted from Ref. [37], Figure 9, 2016, with permission from Elsevier.)

Due to the temperature effect, the transformation hardening is enhanced as the strain rate grows [37]. The driving force in forward transformation progressively increases as the temperature rises. At a strain rate below 10^{-2} s^{-1}, the heat generated by latent heat has to compete with heat conduction and convection in order to raise the temperature. This is similar to the mechanism under monotonic uniaxial loadings.

The number of impact fatigue tests is fewer than that of quasi-static ones. Zurbitu et al. [38] investigated the superelastic repeated-impact behaviors of NiTi SMA wires at a strain rate of 10 s^{-1}. They discovered that the critical transformation stress decreased with increasing cycles in both quasi-static and impact fatigue situations. However, the transformation stress dropped more slowly with cycles in the impact condition since rapid deformation caused a high level of dislocation density which hindered the reduction of martensitic transformation stress. Furthermore, Fitzka et al. [86] found that the intermediate R-phase still occurred in both forward and reverse martensitic transformation when the test frequency was increased to ultrasonic (10^2 s^{-1}). Therefore, the cyclic transformation path is independent of the strain rate.

In summary, the cyclic deformation of NiTi SMA is strongly dependent on the strain rate in the range from 10^{-4} s^{-1} to 10^2 s^{-1}. As the strain rate increases, the superelasticity degenerates more rapidly and the sample temperature increases. The thermo-mechanical coupling effect determines the rate-dependent cyclic behaviors of NiTi SMA.

2.3. Dependence of the Strain Rate Effect on Microstructure

In addition to the loading condition, microstructure also have an important effect on the strain-rate dependent behaviors of NiTi SMAs. For general NiTi SMAs, the strain rate effect of R-phase transformation should be taken into consideration when the total strain is less than 2%. Precipitated phases, such as Ni4Ti3 precipitates, could improve the pseudoelasticity. A smaller grain size of austinite could neutralize the strain rate effect since the temperature effect decreases with a smaller latent heat. Porous SMAs are found with a similar strain rate effect, while SMA composites exhibit excellent impact-resisting performance.

2.3.1. General SMAs

a. R-phase

A rhombohedral (R) phase transformation is much more sensitive to the strain rate compared to martensitic transformation. Since the elongation strain caused by R-phase is usually less than 1% (total martensitic transformation strain is 8%), the R-phase effect can be ignored in most situations. However, when the total strain is less than 2%, the influence of R-phase is worth considering carefully. Helbert et al. [39] built a three-phase pseudodiagram and took R-phase transformation into consideration in explaining the strain effect on sensitivity of NiTi SMA wires. They found that the temperature sensitiveness of R-phase transformation stress was more than 10 MPa/K, which was approximately twice the sensitiveness of martensite phase, as shown in Figure 10. As the strain rate influences the temperature, R-phase transformation stress increases with the strain rate more rapidly than that of martensitic transformation.

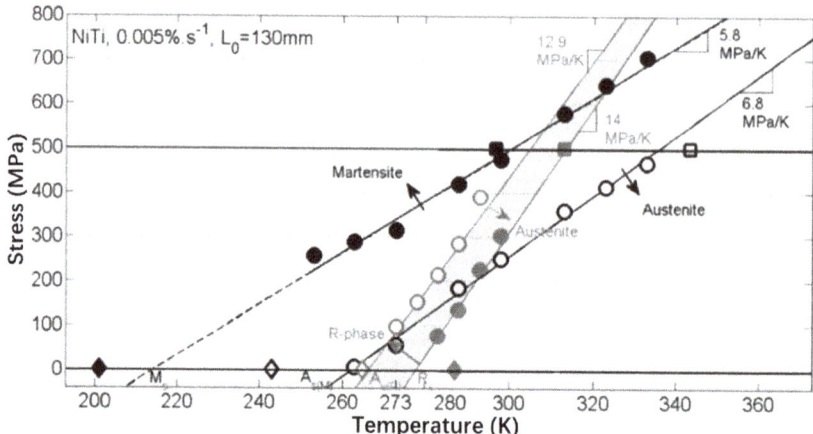

Figure 10. Pseudo-diagram of the studied NiTi alloy. The stress-temperature slope of R-phase is twice that of martensite phase. [39]. (Reprinted from Ref. [39], Figure 7, 2014, with permission from Elsevier.)

b. Precipitated phase

The influence of precipitates on the transformation behaviors of Ni-rich NiTi SMAs has been investigated by a large number of researchers [42,43,87–89]. The precipitated phase of Ni-rich NiTi SMAs is highly dependent on the aging temperature and time, among which the most studied are Ni_4Ti_3 precipitates. The Ni_4Ti_3 precipitates usually introduce R-phase transformation and result in a multistage transformation behavior [87]. Experimental results have shown that the precipitate influence varies with the strain rate.

The precipitation evolution and transformation behavior at quasi-static strain rates were studied and characterized by Fan et al. [42]. The critical transformation stress found was mainly determined by the magnitude of martensitic transformation temperature rather than the appearance of precipitates after aging treatment. Generally, the transformation temperature increases with decreasing aging temperature and increasing aging time.

A recent study by Yu et al. [43] showed that Ni4Ti3 precipitates could improve the pseudoelasticity of NiTi SMAs under impact loading. The critical transformation stress increases with increasing size and volume fraction of precipitates. The best performance in strength is found in the sample with the precipitates dispersed homogeneously within the grains. However, a more comprehensive study on the dependence of the strain rate effect on the precipitated phase is still needed.

c. Grain size

Smaller grains typically reduce the rate-sensitivity of the transformation behavior. Ahadi and Sun [40] studied the rate-dependence of NiTi SMAs on the grain size systematically. Weaker rate dependence was found with smaller grain size, i.e., the difference between the transformation stresses at high and low strain rates diminished as the grain size became smaller. As shown in Figure 11, when the grain shrinks from 90 nm to 10 nm, $\Delta\sigma$ decreases from 135 MPa to zero. Such decreased rate dependence can be explained by smaller temperature variations due to the reduction of latent heat.

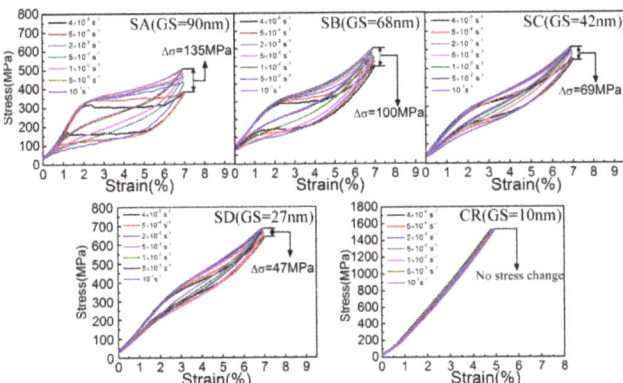

Figure 11. The effect of five Grain Sizes (GS) on the $\sigma - \varepsilon$ curves under monotonic loading–unloading in the strain rate range from $4 \times 10^{-5} \mathrm{s}^{-1}$ to $10^{-1} \mathrm{s}^{-1}$ [40]. (Reprinted from Ref. [40], Figure 5, 2014, with permission from Elsevier.)

2.3.2. Porous SMAs and Composites

The rate dependence of porous SMAs is similar to general solid NiTi SMAs [25], as greater transformation stress is found at higher strain rates.

SMA composites commonly have lamellar structures with NiTi SMAs embedded and exhibit excellent impact-resisting performance [41]. For example, Pappadà et al. dropped a heavy ball on a composite plate to test the impact effect around the strain rate of $10 \mathrm{~s}^{-1}$. They compared the SMA-embedded and steel-embedded plates and showed that the SMA-embedded composites had a better performance in absorbing the impact energy.

3. Models

Typically, the stress increases gradually during transformation, and this transformation hardening is enhanced by increasing the strain rate. In those models that did not directly consider the strain rate effect [90–93], the enhanced transformation hardening was modeled by a phenomenological hardening function with parameters fitted for different strain rates rather than a physical model that inherently considers the hardening mechanism. The hardening function provided extra resistance to transformation, and was generally described as a function of internal state variables with temperature-dependent parameters. These parameters were calibrated by experiments under a specific strain rate and needed to be updated when the strain rate changed. As a result, these models could only give accurate predictions in cases of limited change of the strain rate with a specific group of parameters.

To directly account for the strain rate effects discussed in Section 2, thermal source models are proposed for modeling the NiTi SMA behaviors at low and medium strain rates, and as extended versions, thermal kinetic models are proposed for high strain rates to consider the kinetic effect in shock conditions. These models that take account of the self-heating mechanism and represent the strain rate effect by means of thermal sources can be termed as thermal source models. This is because the strain rate influences the martensitic transformation rate, in turn influences the heat production rate and temperature field. In shock conditions, the velocity of dislocations and phase interfaces set in, so the thermal source model has been extended to thermal kinetic models to consider these kinetic effects.

3.1. Thermal Source Models

At low and medium strain rates, the strain rate effect of NiTi SMA is mainly attributed to temperature variations during transformation, which can be modeled by means of thermal sources in the material [44,45]. The thermal source releases heat in the forward martensitic transformation and absorbs heat in the reverse. The temperature field is hence influenced by a transformation rate that scales with the strain rate.

There are two basic ways to develop the temperature evolution equation in thermal source models. The first method is to directly add external thermal sources to the energy equation, including the latent heat, dissipation heat, and elastic heat [46,47]. The specific form of the thermal source can be constructed empirically. For instance, the released latent heat rate can be represented by a function of the rate of change in a martensitic volume fraction, while the dissipation heat is calculated by a fixed proportion (e.g., 90%) of the total mechanical dissipation energy [48,49].

The second method is to derive the energy equation from an explicit thermodynamic potential with added energy terms related to thermal sources [94,95]. The thermodynamic potential can be either Gibbs or Helmholtz free energy, which is constructed from physical or phenomenological considerations as a function of stress (or strain), temperature, and a set of internal state variables. The chosen form of the free energy should contain the thermal effect introduced by the latent heat, dissipation heat, etc. The evolution equations are then established following a standard thermodynamic procedure.

Two methods lead to similar temperature evolution equations that contain terms with the same physical origins, and both methods need extra constitutive equations for internal state variables [44–47]. The first method that directly adds thermal sources to the energy equation seems to be more simple and easier to realize; however, the second potential method is more favored among researchers in view of its thermodynamic consistency. Therefore, our review will focus on this potential method in explaining the thermal source model. Details of the potential method will be presented in the perspective of thermodynamic theory in Section 3.1.1, followed by a discussion on thermal source components in Section 3.1.2. Simulation examples based on the thermal source model will be shown in Section 3.1.3.

3.1.1. Framework of the Potential Method and the Temperature Evolution Equation

Taking the Gibbs free energy (G) for example, the free energy is usually expressed as a function of a stress tensor S, a temperature T, and other internal variables:

$$G = G(S, T, \gamma^i), \qquad (1)$$

where γ^i represents the ith internal variable. The derivative of Gibbs free energy can then be written by:

$$\dot{G} = \frac{\partial G}{\partial S} : \dot{S} + \frac{\partial G}{\partial T}\dot{T} + \frac{\partial G}{\partial \gamma^i} : \dot{\gamma}^i, \qquad (2)$$

where the Einstein summation convention is assumed. Three conjugate relationships from the second law of thermodynamics are applied:

$$E = -\rho\frac{\partial G}{\partial S}, \qquad (3)$$

$$s = -\frac{\partial G}{\partial T}, \qquad (4)$$

$$\pi^i = -\rho\frac{\partial G}{\partial \gamma^i}, \qquad (5)$$

where E is the strain tensor, ρ is the density, s is the entropy, and π^i is the thermodynamic driving force conjugated to γ^i. Thus, Equation (2) can be simplified to:

$$\dot{G} = -s\dot{T} - \frac{1}{\rho}E : \dot{S} - \frac{1}{\rho}\pi^i : \dot{\gamma}^i. \qquad (6)$$

The first law of thermodynamics can be express as

$$\rho\dot{u} = S : \dot{E} + \rho r - \nabla \cdot q, \qquad (7)$$

where u is the specific internal energy, r is the external heat source density, and q is the heat flux. Note that the mechanical dissipation is well considered in Equation (7). The expression of the internal energy rate is necessary for acquiring the progress of temperature change, which can be obtained from a Legendre transformation:

$$\dot{u} = \dot{G} + T\dot{s} + s\dot{T} + \frac{1}{\rho}\left(S:\dot{E} + E:\dot{S}\right). \tag{8}$$

Substitute (6) into (8), two terms can be removed:

$$\dot{u} = T\dot{s} - \frac{1}{\rho}\pi^i:\dot{\gamma}^i + \frac{1}{\rho}S:\dot{E}. \tag{9}$$

Equations (6), (7), and (9) suggest:

$$-\rho T \frac{\partial}{\partial t}\left(\frac{\partial G}{\partial T}\right) = \pi^i:\dot{\gamma}^i + \rho r - \nabla \cdot q, \tag{10}$$

The form of the Gibbs free energy is not unique. If the internal state variables are selected to be the martensitic volume fraction ξ and the transformation strain ε^t, as used by Boyd and Lagoudas [94], the explicit form of the Gibbs free energy (G_L) can then be defined as:

$$G_L(S, T, \xi, \varepsilon^t) = -\frac{1}{2\rho}S:C:S - \frac{1}{\rho}S:\left[\alpha(T-T_0) + \varepsilon^t\right] + c\left[(T-T_0) - T\ln\left(\frac{T}{T_0}\right)\right] \\ -s_0 T + u_0 + \frac{1}{\rho}f(\xi), \tag{11}$$

where C is the effective compliance tensor, α is the effective thermal expansion coefficient tensor, T_0 is a reference temperature, c is the effective specific heat capacity, s_0 is the effective specific entropy at the reference state, u_0 is the effective specific internal energy at reference state, and $f(\xi)$ is a transformation hardening function.

The effective parameters in (11) are determined by terms of the properties for the pure phases. For instance, the effective thermal expansion coefficient is defined as:

$$\alpha(\xi) = \alpha^A + \xi\left(\alpha^M - \alpha^A\right) = \alpha^A + \xi\Delta\alpha, \tag{12}$$

where the superscripts A and M represent pure austenite and martensite, respectively.

With the assumption that the martensitic transformation happens with no martensitic variant reorientation, the evolution of the transformation strain is then postulated as

$$\dot{\varepsilon}^t = \Lambda\dot{\xi}, \tag{13}$$

where Λ is the transformation tensor. Assume that the specific heat and the thermal expansion coefficients of the two phases are identical:

$$\frac{\partial G_L}{\partial T} = -\frac{1}{\rho}S:\alpha - c\ln\left(\frac{T}{T_0}\right) - s_0, \tag{14}$$

and the driving force π^ξ can be evaluated by:

$$\pi^\xi = S:\Lambda + \frac{1}{2}S:\Delta C:S + \rho\Delta s_0 T - \rho\Delta u_0 - \frac{\partial f}{\partial \xi}. \tag{15}$$

Substitute (14) and (15) into (10):

$$\rho c \dot{T} = -\rho\Delta s_0 T\dot{\xi} + \pi^\xi \dot{\xi} - \alpha:\dot{S}T + \rho r - \nabla \cdot q, \tag{16}$$

which is the governing equation of temperature based on the proposed form of Gibbs free energy in (11).

3.1.2. Components Related to the Thermal Sources

Each term in the temperature evolution Equation (16) corresponds to a specific thermal source, which includes the latent heat, irreversible dissipation heat, elastic heat, heat flux, and external heat sources. These heat components will be discussed individually below.

a. Latent heat

The temperature variations of NiTi SMAs are mainly caused by the latent heat during transformation. Previous phenomenological models assumed that the absorbed or released rate of latent heat had a linear relationship with the change rate of transformation strain [44], while recent models tended to assume a linear relationship with the change rate of martensitic volume fraction [96–98]. For example, the term $\rho \Delta s_0 T \dot{\xi}$ in (16) corresponds to the latent heat, which is determined by the difference of entropy $\rho \Delta s_0$ between two phases, the temperature T, and the martensitic volume fraction change rate $\dot{\xi}$ [99,100].

b. Irreversible dissipation heat

The influence of the dissipation heat on the temperature evolution is second only to that of the latent heat. The mechanism of dissipation of heat are complex, which usually include dissipation by martensitic transformation, martensite reorientation and detwinning, transformation-induced plasticity, and structural plasticity due to the increasing density of dislocations [60,97]. The specific dissipation process depends on the microstructure and external loading conditions.

The contribution of the dissipation heat depends on the strain rate. When the strain rate is low, the dissipation heat by the martensitic transformation in NiTi SMA is smaller than the latent heat by an order of magnitude, which only needs a simple approximation [49] and sometimes can even be ignored [99,101]. In contrast, at medium and high strain rates the dissipation heat could reach a high proportion of the total heat [102,103].

The irreversible dissipation rate can be modeled by a sum of product terms of the thermodynamic driving forces and the change rates of corresponding internal state variables. In general, the thermodynamic driving force is assumed to reach a critical value before the corresponding dissipation process initializes, and then remains a constant during the dissipation. The evolutions of internal state variables are governed by the driving forces during transformation, and they eventually determine the dissipation rate in cases of constant driving forces.

In the potential method, the free energy form is modified to take account of the corresponding dissipation heat in the temperature evolution equation. These modifications basically address the relevant dissipation mechanisms mentioned above. Examples include: (1) The dissipation heat by martensitic transformation as considered by Boyd and Lagoudas [94] in the Gibbs free energy (11) with the corresponding term $\pi^{\xi} \dot{\xi}$ in the temperature evolution equation (3–16). (2) The dissipation heat by the martensite reorientation was taken into account by Šittner et al. [60] in the Gibbs free energy with a function of the volume fraction of each martensite variant. (3) The dissipation heat by transformation-induced plasticity was modeled by Xu et al. [104] in the Gibbs free energy with a term depending on the plastic strain accumulated during transformation. (4) The dissipation heat by general dislocation plasticity was taken into consideration by Heller et al. [105] in the free energy with a term depending on the elevated stress subject to a yield criterion.

c. Elastic heat

The elastic heat rate here is referred to as the power of stress owing to the thermal expansion, which is the term $\alpha : \dot{S} T$ in (16). However, the elastic heat is often neglected in most calculations for its insignificant amount compared to the latent heat [100].

d. Heat flux and external heat source

The heat flux and external heat source, as indicated by the term $\rho r - \nabla \cdot q$ on the right-hand side in (16), correspond to the surface heat conduction/convection and bulk heat production. The heat release and production rates due to external factors also have important influences on the temperature evolution. For example, water-enclosed NiTi SMA wires more readily dissipate heat than air-enclosed ones, and the power of the electricity current greatly influences the temperature of current-driven NiTi SMA wires [5].

3.1.3. Simulation Results with the Thermal Source Model

Quite a few simulations of NiTi SMAs have been carried out under isothermal (quasi-static) or adiabatic conditions based on the thermal source models [50,51,54,55]. Simulation examples have been selected and arranged in the following paragraphs to show the strength of thermal source models in capturing the strain rate effect on both macro- and micro-behaviors of NiTi SMAs. The first simulation case shows the ability of thermal source models to simulate the stress-strain curves with transformation hardening when the strain rate increases. The stress response is attributed to the self-heating mechanism, so the next simulation example investigates the strain rate effect on the temperature variation. Following this, several simulations are presented in studying the contributions of thermal-source components to the temperature variation under various strain rates. The last simulation example discusses the strain rate effect on the nucleation and propagation of phase transformations.

The rate responses of the stress-strain curves are one of the major concerns in simulating the thermomechanical behaviors of NiTi SMAs. Thermal source models can catch the transformation hardening process in the stress-strain curves owing to an increase in strain rate. Simulation examples of isothermal and adiabatic deformation performed by Wang et al. [95] are shown in Figure 12. The simulated stress-strain curves were compared with experimental data in Figure 12a. In the isothermal situation ($\dot{\varepsilon} = 4 \times 10^{-5} \mathrm{s}^{-1}$), the stress stayed constant during transformation; while in the approximately-adiabatic situation ($\dot{\varepsilon} = 4 \times 10^{-2} \mathrm{s}^{-1}$), the stress increased with increasing strain exhibiting a transformation hardening process.

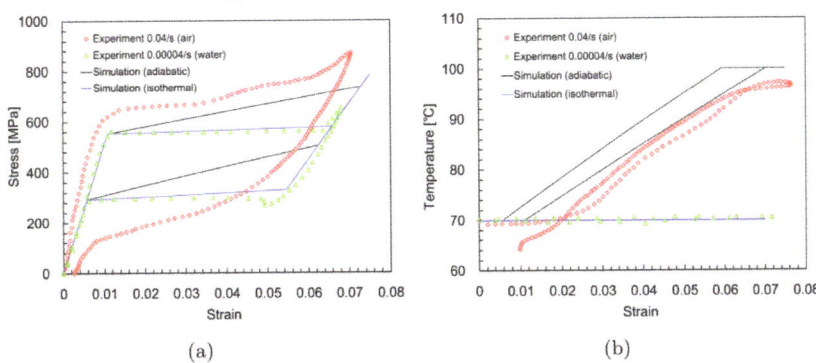

Figure 12. Comparisons between simulation results and experimental data in isothermal and adiabatic conditions: (**a**) stress-strain curves and (**b**) temperature-strain curves [95]. (Reprinted from Ref. [95], Figure 7, 2017, with permission from Institute of Physics Publishing, Ltd.)

The increase of stress during transformation is caused by temperature change. The corresponding simulated temperature-strain curves were compared with experimental data in Figure 12b. The temperature remained constant during transformation in the isothermal situation, while the temperature increased with increasing strain in the approximately adiabatic situation. The strain rate effect on the temperature evolution compares well with experiments as the self-heating mechanism is accounted for in thermal source models.

The influences of the thermal-source components on the temperature change vary with strain rate. The contribution of the dissipation heat component to the temperature rise was investigated under various strain rates [99,101]. At low strain rates, the dissipation heat only had a small effect on the specimen temperature compared to the latent heat, so the dissipation accumulation, calculated by the mechanical dissipation in one cycle, was carefully studied. Grandi et al. [51] measured the areas of the hysteresis cycles in the simulated stress-strain curves and found that the dissipation accumulation increased with increasing strain rate first and then decreased after the peak. The non-monotone trend of the dissipated energy with the strain rate was supported by the experimental results obtained by Zhang et al. [31]. However, the contribution of the dissipation heat to the temperature rise approached that of the latent heat when the strain rate grew above 10^2 s^{-1}. The simulations performed by Shen and Liu [103] showed that the temperature rise increased with increasing strain rate, and the percentage of the dissipation heat ascended to 47% at a strain rate of 1600 s^{-1}, as shown in Figure 13.

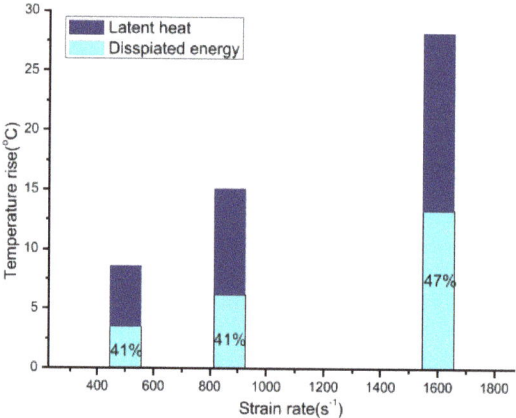

Figure 13. Latent heat and dissipation heat effect on the temperature change [103]. (Reprinted from Ref. [103], Figure 9, 2019, with permission from the publisher Taylor & Francis Ltd, http://www.tandfonline.com).

In addition to the influences of the dissipation heat, the influences of the heat flux component on the temperature change were also explored at different strain rates. For instance, the trends of the maximum temperature rise with the strain rate were studied by Grandi et al. [51] under three different heat transfer coefficients, as shown in Figure 14. Either a low heat transfer coefficient or a high strain rate resulted in a rise in the specimen temperature. This demonstrated the equivalence of the thermal effects caused by decreasing the heat transfer coefficient and increasing the strain rate.

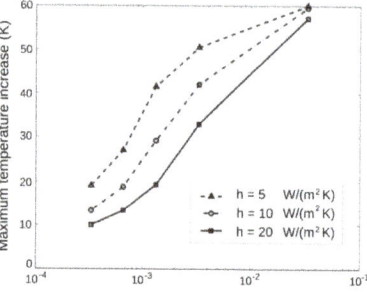

Figure 14. The maximum temperature increase under three different heat transfer coefficients (h) [51]. (Reprinted from Ref. [51], Figure 11, 2012, with permission from Elsevier.)

The strain rate effect on the nucleation and propagation of phase transformation was also captured by thermal source models. Ahmadian et al. [100] simulated the transformation process at strain rates ranging from 10^{-4} s^{-1} to 10^{-1} s^{-1} showing that the number of martensite bands increased as the strain rate increased. The martensitic transformation domains in simulations propagated and widened in a parallel mode. These features are similar to those in experiments (Figure 15).

Figure 15. Phase evolution contours under 3.3×10^{-2}s^{-1} and 1.1×10^{-1}s^{-1} [100]. (Adapted from Ref. [100], Figure 20, 2015, with permission from Elsevier.)

In conclusion, thermal source models are capable of describing the thermomechanical behaviors of NiTi SMAs at low and medium strain rates. However, these models do not consider the kinetic effect, which is one of the main causes for the strain rate effect in dynamic loading conditions. The thermal kinetic model has hence been developed and will be discussed in the next section.

3.2. Thermal Kinetic Models

Thermal source models are not able to capture the sudden rise of stress under dynamic loading conditions. It is observed in experiments that the overall stress level of NiTi SMA increases dramatically when the strain rate increases above 10^3 s^{-1}. In contrast with the rapid growth of stress, the temperature rise reaches a saturation value as the heat by transformation is fully released in the adiabatic process. Thus, the self-heating mechanism can only partially explain the large flow stress in the high-strain-rate deformation.

The rise of overall stress, as well as transformation stress, can be ascribed to the dislocation drag mechanism in the plastic deformation around the phase interface [73,74], as discussed in Section 2. The interface between the austenite and martensite phases contains dislocations that need a much higher driving stress at high strain rates due to the phono drag effect. As a result, the resistance of the phase interface increases sharply leading to a rapid increase in transformation stress. The kinetic properties of the phase interface are therefore of fundamental importance in explaining the great increase in stress during shock conditions.

Some of the earliest phenomenological models were developed by simply introducing a strain rate term into the thermal source model. For instance, Hiroyuki et al. [106] and Auricchio et al. [107] constructed rate-dependent models with thermal sources containing strain rate terms to account for the sole strain-rate effect. Though their simulation results matched well with the experimental results at medium strain rates, their models could hardly reproduce the significant stress rise in dynamic conditions due to the neglect of the kinetic relationship at the phase interface.

Yu et al. [56–58] extended the thermal source model to thermal kinetic model by considering both the self-heating mechanism and kinetic relationship at high strain rates. Based on the thermal source model by Hartl et al. [108], Yu et al. added a term in the traditional transformation driving force to describe the global resistance force of the phase front on transformation. The added resistance force incorporated the velocity of the phase front, and was derived from calculating the needed energy for the kinetic energy change during transformation. The large flow stress and the strain rate effect in the dynamic loading conditions were eventually well-predicted by the thermal kinetic model.

Our review will focus on Yu's model in explaining the thermal kinetic model in view of the limited number of rate-dependent models for dynamic deformation of NiTi SMAs. Necessary numerical verifications based on Yu's model will be presented at the end.

On the basis of the thermal source model proposed by Hartl et al. [108], Yu's model assumes that the dislocation drag effect on the phase interface can be modeled by adding a resistance term to the driving force of the martensitic transformation:

$$\pi'_{tr} = \pi_{tr} - f_D(\xi), \tag{17}$$

where π_{tr} is the driving force in the original quasi-static thermal source model and $f_D(\xi)$ is the added resistance force defined as a linear function of the volume fraction of martensite ξ:

$$f_D(\xi) = K\xi, \tag{18}$$

where K is a constant parameter and described the kinetics related to the strain rate. This parameter can be derived from calculating the kinetic energy change in the wave equation and its form is given as:

$$K = \frac{1}{4}\rho_0 \kappa^2 g_{tr}^2 (\dot{\varepsilon})^2, \tag{19}$$

where g_{tr} is the complete (or maximum) transformation strain, and κ is material parameter which describes the relationship between the strain rate $\dot{\varepsilon}$ and phase boundary velocity C_P as:

$$C_P = \kappa \dot{\varepsilon}. \tag{20}$$

When the strain rate decreases to quasi-static, the resistance force decreases rapidly and finally the thermal kinetic model reduces back to the thermal source model.

Yu and Young [57] then extended the model to three dimensions and applied the thermal kinetic model to simulate the energy band evolutions under high strain rates by the finite element method [58]. The stress-strain curves from experiment data [27] and the thermal kinetic model at five different high strain rates are shown in Figure 16. The critical stress for forward phase transformation increased with increasing strain rate and the hysteresis area shrank when the strain rate was above 9000 s^{-1}. In addition to the experiment results by Guo et al., the kinetic models can match well with other austenitic SMAs.

Compared to the thermal source model, the thermal kinetic model considers both the self-heating effect and the kinetics of the phase interface in dynamic loading. However, the resistance force on the phase interface is estimated globally by a calculation of kinetic energy change, and therefore more effort is still needed to improve the thermal kinetic model in order to consider the localized microstructure of the phase front.

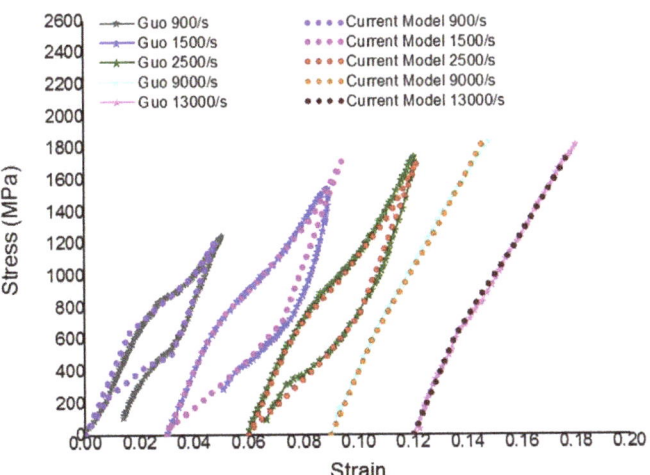

Figure 16. Comparison of stress-strain curves between experiments [27] and simulations [56] at different high strain rates. (Reprinted from Ref. [56], Figure 6, 2017, with permission from Elsevier.)

4. Final Remarks

This paper has reviewed experimental results and constitutive models for the strain rate effect of NiTi SMAs from quasi-static to dynamic loading conditions. An attempt has been made to summarize the physical mechanisms, experimental observations, and models relevant to different strain rates, as shown in Table 1 for uniaxial loading conditions.

Most experimental results reviewed in this paper were under uniaxial loads, while those under shear, indentation, and cyclic loading have also been discussed. Experiments in shear and indentation exhibit similar behaviors to those in uniaxial; as for cyclic loading, superelasticity degeneration and temperature variations are enhanced by increasing the strain rate. The microstructure features such as appearance of the R-phase and precipitated phase, and grain size, also have influences on the strain-rate responses of general NiTi SMAs.

Rate-dependent constitutive models of NiTi SMAs have been built based on the physical mechanisms under different strain rates. Thermal source models have been developed for low and medium strain rates where the strain rate effect could be modeled as thermal sources working in the energy equation. Thermal kinetic models have been extended from thermal source models to consider the kinetic relationship at high strain rates. Both models are effective in modeling the thermodynamic behaviors of NiTi SMAs under corresponding strain rates.

In conclusion, new information provided by this analysis includes; (1) a general plot of the martensitic transformation stress and the austenite yield stress as a function of the strain rate in Figure 2, (2) categorizations of theoretical models based on the physical origins, and (3) a summary of connections between experimental observations, mechanisms, and models for NiTi SMAs at different strain rates in Table 1.

In addition to the information analyzed and summarized above, future research directions are suggested in the following three aspects: (1) new devices and experimental methods for maintaining a stable strain rate in the medium range; (2) comprehensive studies on the influences of microstructure on the strain rate effect; (3) improvements on thermal kinetic models to take account of the localized microstructure of the phase front.

Table 1. A summary of mechanisms, experimental observations, and models at different strain rates.

Strain Rate		Mechanisms	Experimental Observations (Uniaxial Loading, with Increasing Strain Rate)				Models
Range	Rank		Temperature Rise	Transformation Stress	Transformation Mode		
$< 10^{-4} s^{-1}$	Very low	Isothermal process; heat by transformation dissipates to the environment sufficiently	Unnoticeable	Nearly unchanged	Martensite nucleates at few locations and then extends to all fields		Thermal source
$10^{-4} s^{-1} \sim 10^{-1} s^{-1}$	Low	Transition from isothermal to adiabatic; part of heat by transformation is left in the specimen	Increases	Increases gradually	Number of nucleation sites increases and transformation domains propagate in a parallel mode		Thermal source
$10^{-1} s^{-1} \sim 10^{3} s^{-1}$	Medium	Adiabatic process; heat by transformation is fully applied to warm up the specimen	Reaches the maximum	Nearly unchanged	Large martensite zones nucleate at loading ends and spread towards the specimen center		Thermal source
$10^{3} s^{-1} \sim 10^{4} s^{-1}$	High	Adiabatic process; resistance of the phase interface increases dramatically due to the drag effect	Reaches the maximum	Increases dramatically	-		Thermal kinetic
$> 10^{4} s^{-1}$	Very high	Adiabatic process; austenite yields without transformation	-	-	-		Plasticity

Author Contributions: Conceptualization: Z.W. and Y.S.; Methodology: Z.W. and J.L.; Validation: J.L. and Y.S.; Formal analysis: Z.W., J.L., and Y.S.; Investigation: Z.W.; Writing—original draft preparation: Z.W. and J.L.; Writing—review and editing: J.L., Y.S., W.K., G.L., and M.J.; Supervision: Y.S. and X.J.; Project administration: Y.S. and X.J.; Funding acquisition: Y.S. All authors have read and agreed to the published version of the manuscript.

Funding: This work was supported by the National Natural Science Foundation of China (No. 51801122, No. 52071210), the Science and Technology Commission of Shanghai (No. 21ZR1430800).

Data Availability Statement: Not applicable.

Conflicts of Interest: The authors declare no conflict of interest.

References

1. Tušek, J.; Engelbrecht, K.; Mikkelsen, L.P.; Pryds, N. Elastocaloric effect of Ni-Ti wire for application in a cooling device. *J. Appl. Phys.* **2015**, *117*, 124901. [CrossRef]
2. Salvado, F.C.; Teixeira-Dias, F.; Walley, S.M.; Lea, L.J.; Cardoso, J.B. A review on the strain rate dependency of the dynamic viscoplastic response of FCC metals. *Prog. Mater. Sci.* **2017**, *88*, 186–231. [CrossRef]
3. Zhou, M.; Li, Y.; Zhang, C.; Li, S.; Wu, E.; Li, W.; Li, L. The elastocaloric effect of Ni50.8Ti49.2shape memory alloys. *J. Phys. D Appl. Phys.* **2018**, *51*, 135303. [CrossRef]
4. Chien, P.Y.; Martins, J.N.R.; Walsh, L.J.; Peters, O.A. Mechanical and Metallurgical Characterization of Nickel-Titanium Wire Types for Rotary Endodontic Instrument Manufacture. *Materials* **2022**, *15*, 8367. [CrossRef] [PubMed]
5. Casagrande, L.; Menna, C.; Asprone, D.; Ferraioli, M.; Auricchio, F. Chapter 21—Buildings. In *Shape Memory Alloy Engineering*, 2nd ed.; Concilio, A., Antonucci, V., Auricchio, F., Lecce, L., Sacco, E., Eds.; Butterworth-Heinemann: Boston, MA, USA, 2021; pp. 689–729. [CrossRef]
6. Corbi, O. Shape memory alloys and their application in structural oscillations attenuation. *Simul. Model. Pract. Theory* **2003**, *11*, 387–402. [CrossRef]
7. Song, G.; Ma, N.; Li, H.N. Applications of shape memory alloys in civil structures. *Eng. Struct.* **2006**, *28*, 1266–1274. [CrossRef]
8. Torra, V.; Isalgue, A.; Martorell, F.; Terriault, P.; Lovey, F.C. Built in dampers for family homes via SMA: An ANSYS computation scheme based on mesoscopic and microscopic experimental analyses. *Eng. Struct.* **2007**, *29*, 1889–1902. [CrossRef]
9. Dieng, L.; Helbert, G.; Chirani, S.A.; Lecompte, T.; Pilvin, P. Use of Shape Memory Alloys damper device to mitigate vibration amplitudes of bridge cables. *Eng. Struct.* **2013**, *56*, 1547–1556. [CrossRef]
10. Torra, V.; Auguet, C.; Isalgue, A.; Carreras, G.; Terriault, P.; Lovey, F.C. Built in dampers for stayed cables in bridges via SMA. The SMARTeR-ESF project: A mesoscopic and macroscopic experimental analysis with numerical simulations. *Eng. Struct.* **2013**, *49*, 43–57. [CrossRef]
11. Shaw, J.A.; Kyriakides, S. Thermomechanical aspects of NiTi. *J. Mech. Phys. Solids* **1995**, *43*, 1243–1281. [CrossRef]
12. Lin, P.-h.; Tobushi, H.; Tanaka, K.; Hattori, T.; Ikai, A. Influence of strain rate on deformation properties of TiNi shape memory alloy. *JSME Int. J. Ser. A Mech. Mater. Eng.* **1996**, *39*, 117–123. [CrossRef]
13. Tobushi, H.; Shimeno, Y.; Hachisuka, T.; Tanaka, K. Influence of strain rate on superelastic properties of TiNi shape memory alloy. *Mech. Mater.* **1998**, *30*, 141–150. [CrossRef]
14. Tobushi, H.; Okumura, K.; Endo, M.; Tanaka, K. Deformation behavior of TiNi shape memory alloy under strain- or stress-controlled conditions. In *Smart Structures and Materials 2002: Active Materials: Behavior and Mechanics, Proceedings of the SPIE's 9th Annual International Symposium on Smart Structures and Materials, San Diego, CA, USA, 11 July 2002*; Society of Photo-Optical Instrumentation Engineers (SPIE): Bellingham, WA, USA, 2002; pp. 374–385.
15. Bruhns, O.T. Some Remarks on Rate-Sensitivity of NiTi Shape Memory Alloys. *Int. J. Mod. Phys. B* **2008**, *22*, 5406–5412. [CrossRef]
16. Grabe, C.; Bruhns, O.T. On the viscous and strain rate dependent behavior of polycrystalline NiTi. *Int. J. Solids Struct.* **2008**, *45*, 1876–1895. [CrossRef]
17. Chen, W.W.; Wu, Q.; Kang, J.H.; Winfree, N.A. Compressive superelastic behavior of a NiTi shape memory alloy at strain rates of 0.001–750 s^{-1}. *Int. J. Solids Struct.* **2001**, *38*, 8989–8998. [CrossRef]
18. Nemat-Nasser, S.; Choi, J.Y. Thermomechanical response of an Ni–Ti–Cr shape-memory alloy at low and high strain rates. *Philos. Mag.* **2006**, *86*, 1173–1187. [CrossRef]
19. Shen, L.; Liu, Y.; Hui, M. Dynamic thermo-mechanical behaviors of SME TiNi alloys subjected to shock loading. *Acta Mech. Sin.* **2020**, *36*, 1336–1349. [CrossRef]
20. Zurbitu, J.; Castillo, G.; Urrutibeascoa, I.; Aurrekoetxea, J. Low-energy tensile-impact behavior of superelastic NiTi shape memory alloy wires. *Mech. Mater.* **2009**, *41*, 1050–1058. [CrossRef]
21. Zurbitu, J.; Kustov, S.; Castillo, G.; Aretxabaleta, L.; Cesari, E.; Aurrekoetxea, J. Instrumented tensile–impact test method for shape memory alloy wires. *Mater. Sci. Eng. A* **2009**, *524*, 108–111. [CrossRef]
22. Kocks, U.F.; Argon, A.S.; Ashby, M.F. Thermodynamics and kinetics of slip. *Progr. Mater. Sci.* **1975**, *19*, 308.

23. Nemat-Nasser, S.; Choi, J.-Y.; Guo, W.-G.; Isaacs, J.B. Very high strain-rate response of a NiTi shape-memory alloy. *Mech. Mater.* **2005**, *37*, 287–298. [CrossRef]
24. Nemat-Nasser, S.; Choi, J.Y. Strain rate dependence of deformation mechanisms in a Ni–Ti–Cr shape-memory alloy. *Acta Mater.* **2005**, *53*, 449–454. [CrossRef]
25. Nemat-Nasser, S.; Su, Y.; Guo, W.-G.; Isaacs, J. Experimental characterization and micromechanical modeling of superelastic response of a porous NiTi shape-memory alloy. *J. Mech. Phys. Solids* **2005**, *53*, 2320–2346. [CrossRef]
26. Zhang, X.; Wang, G.; Luo, B.; Bland, S.N.; Tan, F.; Zhao, F.; Zhao, J.; Sun, C.; Liu, C. Mechanical response of near-equiatomic NiTi alloy at dynamic high pressure and strain rate. *J. Alloys Compd.* **2018**, *731*, 569–576. [CrossRef]
27. Guo, W.G.; Su, J.; Su, Y.; Chu, S.Y. On phase transition velocities of NiTi shape memory alloys. *J. Alloys Compd.* **2010**, *501*, 70–76. [CrossRef]
28. Yang, Z.; Wang, H.; Huang, Y.; Ye, X.; Li, J.; Zhang, C.; Li, H.; Pang, B.; Tian, Y.; Huang, C.; et al. Strain rate dependent mechanical response for monoclinic NiTi shape memory alloy: Micromechanical decomposition and model validation via neutron diffraction. *Mater. Des.* **2020**, *191*, 108656. [CrossRef]
29. Gadaj, S.; Nowacki, W.; Pieczyska, E.; Tobushi, H. Temperature measurement as a new technique applied to the phase transformation study in a TiNi shape memory alloy subjected to tension. *Arch. Metall. Mater.* **2005**, *50*, 661–674.
30. Saletti, D.; Pattofatto, S.; Zhao, H. Measurement of phase transformation properties under moderate impact tensile loading in a NiTi alloy. *Mech. Mater.* **2013**, *65*, 1–11. [CrossRef]
31. Zhang, X.; Feng, P.; He, Y.; Yu, T.; Sun, Q. Experimental study on rate dependence of macroscopic domain and stress hysteresis in NiTi shape memory alloy strips. *Int. J. Mech. Sci.* **2010**, *52*, 1660–1670. [CrossRef]
32. Qiu, Y.; Young, M.L.; Nie, X. High Strain Rate Compression of Martensitic NiTi Shape Memory Alloys. *Shape Mem. Superelasticity* **2015**, *1*, 310–318. [CrossRef]
33. Manach, P.-Y.; Favier, D. Shear and tensile thermomechanical behavior of near equiatomic NiTi alloy. *Mater. Sci. Eng. A* **1997**, *222*, 45–57. [CrossRef]
34. Huang, H.; Durand, B.; Sun, Q.P.; Zhao, H. An experimental study of NiTi alloy under shear loading over a large range of strain rates. *Int. J. Impact Eng.* **2017**, *108*, 402–413. [CrossRef]
35. Amini, A.; He, Y.; Sun, Q. Loading rate dependency of maximum nanoindentation depth in nano-grained NiTi shape memory alloy. *Mater. Lett.* **2011**, *65*, 464–466. [CrossRef]
36. Shahirnia, M.; Farhat, Z.; Jarjoura, G. Effects of temperature and loading rate on the deformation characteristics of superelastic TiNi shape memory alloys under localized compressive loads. *Mater. Sci. Eng. A* **2011**, *530*, 628–632. [CrossRef]
37. Kan, Q.; Yu, C.; Kang, G.; Li, J.; Yan, W. Experimental observations on rate-dependent cyclic deformation of super-elastic NiTi shape memory alloy. *Mech. Mater.* **2016**, *97*, 48–58. [CrossRef]
38. Zurbitu, J.; Santamarta, R.; Picornell, C.; Gan, W.M.; Brokmeier, H.G.; Aurrekoetxea, J. Impact fatigue behavior of superelastic NiTi shape memory alloy wires. *Mater. Sci. Eng. A* **2010**, *528*, 764–769. [CrossRef]
39. Helbert, G.; Saint-Sulpice, L.; Arbab Chirani, S.; Dieng, L.; Lecompte, T.; Calloch, S.; Pilvin, P. Experimental characterisation of three-phase NiTi wires under tension. *Mech. Mater.* **2014**, *79*, 85–101. [CrossRef]
40. Ahadi, A.; Sun, Q. Effects of grain size on the rate-dependent thermomechanical responses of nanostructured superelastic NiTi. *Acta Mater.* **2014**, *76*, 186–197. [CrossRef]
41. Pappadà, S.; Rametta, R.; Toia, L.; Coda, A.; Fumagalli, L.; Maffezzoli, A. Embedding of Superelastic SMA Wires into Composite Structures: Evaluation of Impact Properties. *J. Mater. Eng. Perform.* **2009**, *18*, 522–530. [CrossRef]
42. Fan, Q.C.; Zhang, Y.H.; Wang, Y.Y.; Sun, M.Y.; Meng, Y.T.; Huang, S.K.; Wen, Y.H. Influences of transformation behavior and precipitates on the deformation behavior of Ni-rich NiTi alloys. *Mater. Sci. Eng. A* **2017**, *700*, 269–280. [CrossRef]
43. Yu, H.; Qiu, Y.; Young, M.L. Influence of Ni4Ti3 precipitate on pseudoelasticity of austenitic NiTi shape memory alloys deformed at high strain rate. *Mater. Sci. Eng. A* **2021**, *804*, 140753. [CrossRef]
44. Tobushi, H.; Takata, K.; Shimeno, Y.; Nowacki, W.K.; Gadaj, S.P. Influence of strain rate on superelastic behaviour of TiNi shape memory alloy. *J. Mater. Des. Appl.* **1999**, *213*, 93–102. [CrossRef]
45. Seelecke, S. Modeling the dynamic behavior of shape memory alloys. *Int. J. Non-Linear Mech.* **2002**, *37*, 1363–1374. [CrossRef]
46. He, Y.J.; Sun, Q.P. On non-monotonic rate dependence of stress hysteresis of superelastic shape memory alloy bars. *Int. J. Solids Struct.* **2011**, *48*, 1688–1695. [CrossRef]
47. Yang, S.Y.; Dui, G.S. Temperature analysis of one-dimensional NiTi shape memory alloys under different loading rates and boundary conditions. *Int. J. Solids Struct.* **2013**, *50*, 3254–3265. [CrossRef]
48. Azadi, B.; Rajapakse, R.K.N.D.; Maijer, D.M. One-dimensional thermomechanical model for dynamic pseudoelastic response of shape memory alloys. *Smart Mater. Struct.* **2006**, *15*, 996–1008. [CrossRef]
49. Yang, S.; Dui, G.; Liu, B. Modeling of rate-dependent damping capacity of one-dimensional superelastic shape memory alloys. *J. Intell. Mater. Syst. Struct.* **2012**, *24*, 431–440. [CrossRef]
50. Berti, V.; Fabrizio, M.; Grandi, D. Hysteresis and phase transitions for one-dimensional and three-dimensional models in shape memory alloys. *J. Math. Phys.* **2010**, *51*, 062901. [CrossRef]

51. Grandi, D.; Maraldi, M.; Molari, L. A macroscale phase-field model for shape memory alloys with non-isothermal effects: Influence of strain rate and environmental conditions on the mechanical response. *Acta Mater.* **2012**, *60*, 179–191. [CrossRef]
52. Dhote, R.P.; Gomez, H.; Melnik, R.N.V.; Zu, J. Shape memory alloy nanostructures with coupled dynamic thermo-mechanical effects. *Comput. Phys. Commun.* **2015**, *192*, 48–53. [CrossRef]
53. Cui, S.; Wan, J.; Zuo, X.; Chen, N.; Zhang, J.; Rong, Y. Three-dimensional, non-isothermal phase-field modeling of thermally and stress-induced martensitic transformations in shape memory alloys. *Int. J. Solids Struct.* **2017**, *109*, 1–11. [CrossRef]
54. Xie, X.; Kang, G.; Kan, Q.; Yu, C.; Peng, Q. Phase field modeling to transformation induced plasticity in super-elastic NiTi shape memory alloy single crystal. *Model. Simul. Mater. Sci. Eng.* **2019**, *27*, 045001. [CrossRef]
55. Xie, X.; Kang, G.; Kan, Q.; Yu, C. Phase-field theory based finite element simulation on thermo-mechanical cyclic deformation of polycrystalline super-elastic NiTi shape memory alloy. *Comput. Mater. Sci.* **2020**, *184*, 109899. [CrossRef]
56. Yu, H.; Young, M.L. One-dimensional thermomechanical model for high strain rate deformation of austenitic shape memory alloys. *J. Alloys Compd.* **2017**, *710*, 858–868. [CrossRef]
57. Yu, H.; Young, M.L. Three-dimensional modeling for deformation of austenitic NiTi shape memory alloys under high strain rate. *Smart Mater. Struct.* **2018**, *27*, 015031. [CrossRef]
58. Yu, H.; Young, M.L. Effect of temperature on high strain rate deformation of austenitic shape memory alloys by phenomenological modeling. *J. Alloys Compd.* **2019**, *797*, 194–204. [CrossRef]
59. Nnamchi, P.; Younes, A.; González, S. A review on shape memory metallic alloys and their critical stress for twinning. *Intermetallics* **2019**, *105*, 61–78. [CrossRef]
60. Šittner, P.; Sedlák, P.; Seiner, H.; Sedmák, P.; Pilch, J.; Delville, R.; Heller, L.; Kadeřávek, L. On the coupling between martensitic transformation and plasticity in NiTi: Experiments and continuum based modelling. *Prog. Mater. Sci.* **2018**, *98*, 249–298. [CrossRef]
61. Cisse, C.; Zaki, W.; Ben Zineb, T. A review of constitutive models and modeling techniques for shape memory alloys. *Int. J. Plast.* **2016**, *76*, 244–284. [CrossRef]
62. Chen, W.; Song, B. Temperature dependence of a NiTi shape memory alloy's superelastic behavior at a high strain rate. *J. Mech. Mater. Struct.* **2006**, *1*, 339–356. [CrossRef]
63. Adharapurapu, R.R.; Jiang, F.; Vecchio, K.S.; Gray, G.T. Response of NiTi shape memory alloy at high strain rate: A systematic investigation of temperature effects on tension–compression asymmetry. *Acta Mater.* **2006**, *54*, 4609–4620. [CrossRef]
64. Elibol, C.; Wagner, M.F.X. Strain rate effects on the localization of the stress-induced martensitic transformation in pseudoelastic NiTi under uniaxial tension, compression and compression–shear. *Mater. Sci. Eng. A* **2015**, *643*, 194–202. [CrossRef]
65. Nemat-Nasser, S.; Yong Choi, J.; Guo, W.-G.; Isaacs, J.B.; Taya, M. High Strain-Rate, Small Strain Response of a NiTi Shape-Memory Alloy. *J. Eng. Mater. Technol.* **2005**, *127*, 83–89. [CrossRef]
66. Kapoor, R.; Nemat-Nasser, S. Comparison between high and low strain-rate deformation of tantalum. *Metall. Mater. Trans. A* **2000**, *31*, 815–823. [CrossRef]
67. Austin, R.A.; McDowell, D.L. A dislocation-based constitutive model for viscoplastic deformation of fcc metals at very high strain rates. *Int. J. Plast.* **2011**, *27*, 1–24. [CrossRef]
68. Shahba, A.; Ghosh, S. Crystal plasticity FE modeling of Ti alloys for a range of strain-rates. Part I: A unified constitutive model and flow rule. *Int. J. Plast.* **2016**, *87*, 48–68. [CrossRef]
69. Luscher, D.J.; Addessio, F.L.; Cawkwell, M.J.; Ramos, K.J. A dislocation density-based continuum model of the anisotropic shock response of single crystal α-cyclotrimethylene trinitramine. *J. Mech. Phys. Solids* **2017**, *98*, 63–86. [CrossRef]
70. Dayananda, G.N.; Rao, M.S. Effect of strain rate on properties of superelastic NiTi thin wires. *Mater. Sci. Eng. A* **2008**, *486*, 96–103. [CrossRef]
71. Boyce, B.L.; Crenshaw, T.B. *Servohydraulic Methods for Mechanical Testing in the Sub-Hopkinson Rate Regime up to Strain Rates of 500 1/s*; Sandia National Laboratories: Albuquerque, NM, USA, 2005.
72. Xu, R.; Cui, L.; Zheng, Y. The dynamic impact behavior of NiTi alloy. *Mater. Sci. Eng. A* **2006**, *438–440*, 571–574. [CrossRef]
73. Tang, B.; Tang, B.; Han, F.; Yang, G.; Li, J. Influence of strain rate on stress induced martensitic transformation in β solution treated TB8 alloy. *J. Alloys Compd.* **2013**, *565*, 1–5. [CrossRef]
74. Niitsu, K.; Date, H.; Kainuma, R. Thermal activation of stress-induced martensitic transformation in Ni-rich Ti-Ni alloys. *Scr. Mater.* **2020**, *186*, 263–267. [CrossRef]
75. Hudspeth, M.; Sun, T.; Parab, N.; Guo, Z.; Fezzaa, K.; Luo, S.; Chen, W. Simultaneous X-ray diffraction and phase-contrast imaging for investigating material deformation mechanisms during high-rate loading. *J. Synchrotron Radiat.* **2015**, *22*, 49–58. [CrossRef] [PubMed]
76. Wang, M.; Jiang, S.; Zhang, Y. Phase Transformation, Twinning, and Detwinning of NiTi Shape-Memory Alloy Subject to a Shock Wave Based on Molecular-Dynamics Simulation. *Materials* **2018**, *11*, 2334. [CrossRef] [PubMed]
77. Yazdandoost, F.; Mirzaeifar, R. Stress Wave and Phase Transformation Propagation at the Atomistic Scale in NiTi Shape Memory Alloys Subjected to Shock Loadings. *Shape Mem. Superelasticity* **2018**, *4*, 435–449. [CrossRef]
78. Yin, Q.; Wu, X.; Huang, C.; Wang, X.; Wei, Y. Atomistic study of temperature and strain rate-dependent phase transformation behaviour of NiTi shape memory alloy under uniaxial compression. *Philos. Mag.* **2015**, *95*, 2491–2512. [CrossRef]

79. Ko, W.-S.; Maisel, S.B.; Grabowski, B.; Jeon, J.B.; Neugebauer, J. Atomic scale processes of phase transformations in nanocrystalline NiTi shape-memory alloys. *Acta Mater.* **2017**, *123*, 90–101. [CrossRef]
80. Yazdandoost, F.; Sadeghi, O.; Bakhtiari-Nejad, M.; Elnahhas, A.; Shahab, S.; Mirzaeifar, R. Energy dissipation of shock-generated stress waves through phase transformation and plastic deformation in NiTi alloys. *Mech. Mater.* **2019**, *137*, 103090. [CrossRef]
81. Millett, J.C.F.; Bourne, N.K.; Gray, G.T. Behavior of the shape memory alloy NiTi during one-dimensional shock loading. *J. Appl. Phys.* **2002**, *92*, 3107–3110. [CrossRef]
82. Farhat, Z.; Jarjoura, G.; Shahirnia, M. Dent Resistance and Effect of Indentation Loading Rate on Superelastic TiNi Alloy. *Metall. Mater. Trans. A* **2013**, *44*, 3544–3551. [CrossRef]
83. Yu, C.; Kang, G.; Kan, Q. A physical mechanism based constitutive model for temperature-dependent transformation ratchetting of NiTi shape memory alloy: One-dimensional model. *Mech. Mater.* **2014**, *78*, 1–10. [CrossRef]
84. Miyazaki, S. Thermal and Stress Cycling Effects and Fatigue Properties of Ni-Ti Alloys. In *Engineering Aspects of Shape Memory Alloys*; Duerig, T.W., Melton, K.N., Stöckel, D., Wayman, C.M., Eds.; Butterworth-Heinemann: Boston, MA, USA, 1990; pp. 394–413. [CrossRef]
85. Lin, P.H.; Tobushi, H.; Tanaka, K.; Hattori, T.; Makita, M. Pseudoelastic Behaviour of TiNi Shape Memory Alloy Subjected to Strain Variations. *J. Intell. Mater. Syst. Struct.* **1994**, *5*, 694–701. [CrossRef]
86. Fitzka, M.; Rennhofer, H.; Catoor, D.; Reiterer, M.; Lichtenegger, H.; Checchia, S.; di Michiel, M.; Irrasch, D.; Gruenewald, T.A.; Mayer, H. High Speed In Situ Synchrotron Observation of Cyclic Deformation and Phase Transformation of Superelastic Nitinol at Ultrasonic Frequency. *Exp. Mech.* **2019**, *60*, 317–328. [CrossRef]
87. Bragov, A.M.; Danilov, A.N.; Konstantinov, A.Y.; Lomunov, A.K.; Motorin, A.S.; Razov, A.I. Mechanical and structural aspects of high-strain-rate deformation of NiTi alloy. *Phys. Met. Metallogr.* **2015**, *116*, 385–392. [CrossRef]
88. Kaya, I.; Karaca, H.E.; Nagasako, M.; Kainuma, R. Effects of aging temperature and aging time on the mechanism of martensitic transformation in nickel-rich NiTi shape memory alloys. *Mater. Charact.* **2020**, *159*, 110034. [CrossRef]
89. Adharapurapu, R.R.; Jiang, F.; Bingert, J.F.; Vecchio, K.S. Influence of cold work and texture on the high-strain-rate response of Nitinol. *Mater. Sci. Eng. A* **2010**, *527*, 5255–5267. [CrossRef]
90. Liang, C.; Rogers, C.A. One-Dimensional Thermomechanical Constitutive Relations for Shape Memory Materials. *J. Intell. Mater. Syst. Struct.* **1990**, *1*, 207–234. [CrossRef]
91. Boyd, J.G.; Lagoudas, D.C. A thermodynamical constitutive model for shape memory materials. Part I. The monolithic shape memory alloy. *Int. J. Plast.* **1996**, *12*, 805–842. [CrossRef]
92. Boyd, J.G.; Lagoudas, D.C. A thermodynamical constitutive model for shape memory materials. Part II. The SMA composite material. *Int. J. Plast.* **1996**, *12*, 843–873. [CrossRef]
93. Leclercq, S.; Lexcellent, C. A general macroscopic description of the thermomechanical behavior of shape memory alloys. *J. Mech. Phys. Solids* **1996**, *44*, 953–980. [CrossRef]
94. Lagoudas, D.C. *Shape Memory Alloys: Modeling and Engineering Applications*; Springer: New York, NY, USA, 2008.
95. Wang, J.; Moumni, Z.; Zhang, W.; Xu, Y.; Zaki, W. A 3D finite-strain-based constitutive model for shape memory alloys accounting for thermomechanical coupling and martensite reorientation. *Smart Mater. Struct.* **2017**, *26*, 065006. [CrossRef]
96. Kadkhodaei, M.; Rajapakse, R.K.N.D.; Mahzoon, M.; Salimi, M. Modeling of the cyclic thermomechanical response of SMA wires at different strain rates. *Smart Mater. Struct.* **2007**, *16*, 2091–2101. [CrossRef]
97. Sadjadpour, A.; Bhattacharya, K. A micromechanics inspired constitutive model for shape-memory alloys: The one-dimensional case. *Smart Mater. Struct.* **2007**, *16*, S51–S62. [CrossRef]
98. Zhu, S.; Zhang, Y. A thermomechanical constitutive model for superelastic SMA wire with strain-rate dependence. *Smart Mater. Struct.* **2007**, *16*, 1696–1707. [CrossRef]
99. Motahari, S.A.; Ghassemieh, M. Multilinear one-dimensional shape memory material model for use in structural engineering applications. *Eng. Struct.* **2007**, *29*, 904–913. [CrossRef]
100. Ahmadian, H.; Hatefi Ardakani, S.; Mohammadi, S. Strain-rate sensitivity of unstable localized phase transformation phenomenon in shape memory alloys using a non-local model. *Int. J. Solids Struct.* **2015**, *63*, 167–183. [CrossRef]
101. Lagoudas, D.C.; Ravi-Chandar, K.; Sarh, K.; Popov, P. Dynamic loading of polycrystalline shape memory alloy rods. *Mech. Mater.* **2003**, *35*, 689–716. [CrossRef]
102. Liu, Y.; Shan, L.; Shan, J.; Hui, M. Experimental study on temperature evolution and strain rate effect on phase transformation of TiNi shape memory alloy under shock loading. *Int. J. Mech. Sci.* **2019**, *156*, 342–354. [CrossRef]
103. Shen, L.; Liu, Y. Temperature evolution associated with dynamic phase transformation in shape-memory TiNi alloys. *Phase Transit.* **2019**, *92*, 755–771. [CrossRef]
104. Xu, L.; Solomou, A.; Baxevanis, T.; Lagoudas, D. Finite strain constitutive modeling for shape memory alloys considering transformation-induced plasticity and two-way shape memory effect. *Int. J. Solids Struct.* **2021**, *221*, 42–59. [CrossRef]
105. Heller, L.; Šittner, P.; Sedlák, P.; Seiner, H.; Tyc, O.; Kadeřávek, L.; Sedmák, P.; Vronka, M. Beyond the strain recoverability of martensitic transformation in NiTi. *Int. J. Plast.* **2019**, *116*, 232–264. [CrossRef]

106. Hiroyuki, N.; Ying, Z.; Minoru, T.; Weinong, W.C.; Yuta, U.; Shunji, S. Strain-rate effects on TiNi and TiNiCu shape memory alloys. In Proceedings of the SPIE Smart Structures and Materials + Nondestructive Evaluation and Health Monitoring, San Diego, CA, USA, 16 May 2005; pp. 355–363.
107. Auricchio, F.; Fugazza, D.; DesRoches, R. A 1D rate-dependent viscous constitutive model for superelastic shape-memory alloys: Formulation and comparison with experimental data. *Smart Mater. Struct.* **2007**, *16*, S39–S50. [CrossRef]
108. Hartl, D.J.; Chatzigeorgiou, G.; Lagoudas, D.C. Three-dimensional modeling and numerical analysis of rate-dependent irrecoverable deformation in shape memory alloys. *Int. J. Plast.* **2010**, *26*, 1485–1507. [CrossRef]

Disclaimer/Publisher's Note: The statements, opinions and data contained in all publications are solely those of the individual author(s) and contributor(s) and not of MDPI and/or the editor(s). MDPI and/or the editor(s) disclaim responsibility for any injury to people or property resulting from any ideas, methods, instructions or products referred to in the content.

MDPI
St. Alban-Anlage 66
4052 Basel
Switzerland
Tel. +41 61 683 77 34
Fax +41 61 302 89 18
www.mdpi.com

Metals Editorial Office
E-mail: metals@mdpi.com
www.mdpi.com/journal/metals

www.ingramcontent.com/pod-product-compliance
Lightning Source LLC
LaVergne TN
LVHW070655100526
838202LV00013B/972